VIRTUAL EXPERIMENTS
IN FOOD PROCESSING

ii

VIRTUAL EXPERIMENTS
IN FOOD PROCESSING

R. Paul Singh

Professor of Food Engineering
Department of Biological and Agricultural Engineering
Department of Food Science and Technology
University of California
Davis, California

Ferruh Erdogdu

Assistant Professor of Food Engineering
Department of Food Engineering
University of Mersin
Ciftlikkoy-Mersin, Turkey

RAR Press
Davis, California

International Standard Book Number 0-9748638-0-7

Library of Congress Catalog Card Number: 2003099120

RAR Press
2317 Lassen Pl, Davis, California 95616, USA
Email: rps@rpaulsingh.com
http://www.rpaulsingh.com

PRINTED IN THE UNITED STATES OF AMERICA

About the Authors

R. Paul Singh is a professor of food engineering at the University of California, Davis, where he teaches food engineering at undergraduate and graduate level. He is a fellow of the Institute of Food Technologists (IFT), American Society of Agricultural Engineers (ASAE), and the International Academy of Food Science and Technology. He received the *Samuel Cate Prescott Award for Research* (1982), and *International Award* (1988) from IFT, *Young Educator Award* (1986) from ASAE, *Food Engineering Award* (1997) from the International Association of Food Industry Suppliers and the *Grand Prix Research Award* (2001) from the Japanese Food Machinery Manufacturers Association. He is an author or co-author of 15 books and over 200 refereed papers. He frequently consults with the food industry and international agencies and he has contributed to research and educational programs in more than 36 countries. His research involves use of computer-aided methods to study a variety of food operations such as freezing, drying, thawing, frying, and impingement systems to improve product quality and process efficiency.

Ferruh Erdogdu is an assistant professor of food engineering at the University of Mersin, Turkey. He graduated from the Department of Food Engineering at Hacettepe University in Ankara with the highest grades in 1992. In 1994, he succeeded in a nation wide examination by the Ministry of National Education of Turkey to pursue masters and Ph.D. degrees in food engineering in the United States. He received his Master of Engineering degree in 1996 and a Ph.D. in 2000 at the University of Florida, Gainesville. During his studies, he maintained a status of Distinguished Scholar. He received outstanding academic achievement awards from the College of Engineering (from 1997 to 2000) and College of Agriculture (1999) and won the student paper competition of Food Engineering Division, Institute of Food Technologists, in 1997. He conducted his postdoctoral work at the University of California, Davis. In 2001, he joined the faculty of Food Engineering at the University of Mersin, Turkey, where he has been teaching undergraduate and graduate-level courses on topics in food engineering. He is an author or coauthor of more than 15 research papers and more than 30 presentations. He is currently a professional member of the Institute of Food Technologists. He is serving in the editorial board of the Journal of Food Process Engineering, and his current research interests include mathematical modeling of heat and mass transfer operations in food processing.

PREFACE

In this book and the accompanying CD, we have drawn on results from over 30 years of research on selected computer applications in food processing to develop educational materials suitable for teaching. Seventeen virtual experiments have been prepared that may be conducted using the software presented in the CD. The accompanying text provides detailed procedures required to conduct a given virtual experiment. These experiments may be used to augment existing laboratory courses, or as contents of a stand-alone virtual laboratory course in the food science curriculum.

The topics selected for virtual experiments represent major food processes, and in each case an experiment is designed with following components:

First, a collection of multimedia materials including photographs, schematics and animations of process equipment are presented to view industrial practice and laboratory procedures relevant to the experiment. This allows a student to become visually familiar with the industrial practice and experimental procedures used in a laboratory.

The second component involves a process simulation that uses advanced mathematical models to predict physical, chemical or microbiological changes in the food due to the process. These mathematical models have been extensively validated with experimental data in published literature. Therefore, the predictions of the food processes are considered to be highly reliable. The user is shielded from the complexity of the models, because of the consistent user-friendly input/output procedures that have been developed for each virtual experiment.

The third component relates to critical thinking skills required in data analysis. Although the simulation programs may be enhanced to do all the analysis, this would minimize student's learning. Therefore, from each virtual experiment, a student obtains results in the form of spreadsheets. Students are then asked to analyze the data by making required plots, derive important parameters, conduct statistical analysis, and discuss key observations. This part is similar to what is normally done with data obtained from a real-life laboratory experiment.

The fourth component is report writing. Online links to over 60 industrial web sites are provided. These links are included for students to conduct research for data analysis and report writing. Discussion questions are included to prompt students to make key observations while conducting experiments. Another feature in the CD is MyJournal, an electronic text file, where a student may keep notes as the experiments are being conducted. This text file is saved on the student's computer and may be used later in preparing the written report.

We have used these virtual experiments in teaching students enrolled in courses in food processing and engineering. These laboratories have allowed the students to take full advantage of the vast computational power of the modern personal computers and to improve their problem solving skills.

Many graduate students have contributed to the development of these laboratory experiments. In particular, we would like to acknowledge the assistance of Brent Anderson, Maria Anderson, Maria Ferrua, Miguel Santos, Arnab Sarkar, and Daniel Voit.

R. Paul Singh
Ferruh Erdogdu

Contents

xii

Chapter 1 Getting Started – Installation and General Information

1.1 Installation

In the accompanying CD, you will find a file named **Enter_Lab.htm**. By clicking on this file, a screen will appear as shown in Fig. 1.1. Launch any experiment by clicking on its name. To run the programs without the CD, first create a folder on the hard disk. Then copy **Enter_lab.htm** and the folder **LabExperiments** into the new folder on the hard disk. It is important to keep the file structure same as on the CD to maintain correct links. Please refer to the License Agreement (page 125) prior to making any copy.

If the computer you are using is part of a network, then you will need the assistance of a System Administrator to first save certain *.dll and *.ocx files in the System folder of the server. Instructions for the System Administrator are available on the Read_Me.txt file. Once the *.dll and *.ocx files are saved in the system folder of the server, you will be able to run the programs directly from the CD on the network computers. Program files must not be copied on the network server.

If you get the following message during start up,

> *"A file being copied is not newer than the file currently on your system. It is recommended that you keep your existing file ...... Do you want to keep this file?"*

select **YES** option for this message.

Virtual Experiments in Food Processing

Note: To view files in each laboratory experiment, you must have a Flash player installed in the browser. The player is available free of charge from Macromedia.

To view pdf file in "Getting Started," you will require an Adobe® Reader® available free of charge from Acrobat

Getting Started (pdf)

Energy Requirements of Pumping Apple Juice

Rheological Properties of Foods -- Determining Flow Properties of Vanilla Pudding

Temperature sensors -- Response Time of Thermocouples

Convective Heat transfer -- Determining Heat Transfer Coefficient in Air and Water

Heat exchangers - Heating Milk in a Tubular Heat Exchanger

Heating liquid foods -- Heating Tomato Juice in a Steam-Jacketed Kettle

Figure 1.1. A partial list of virtual experiments.

1.2 Startup

Each program runs like any other Windows application starting with a Start-up screen (Fig. 1.2). The Start-up screen loads the program in 2 to 3 seconds, or you may click anywhere on this screen to start the program instantly.

1.3 Menu Bar

In the menu bar of each program (Fig. 1.3), there are four options:

- **Overview**: To open the laboratory overview at any time
- **MyJournal**: To open a Notepad® file to enter any data or to take notes during the experiment
- **Print**: To print the active screen
- **Exit**: To exit the program. Use this option to exit the program. **Caution**: You should stop any running program before using **Exit.**

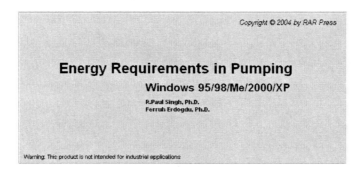

Figure 1.2. Start-up screen for an experiment on energy requirements in pumping.

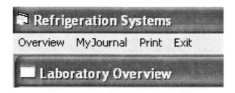

Figure 1.3. The four options of the menu bar (Overview, MyJournal, Print and Exit).

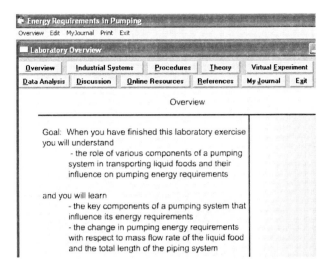

Figure 1.4   Overview screen with tab buttons to access all components of a laboratory exercise.

1.4 Laboratory Overview

When a program is launched, a laboratory overview screen is shown as in Fig. 1.4. It includes several buttons that may be selected to view various components of the laboratory exercise. The function of each button is as follows:

- **Overview**: To view the goal and objectives of the experiment
- **Industrial Systems**: To view the industrial use of equipment related to a given experiment.
- **Procedures**: To view key procedural steps that are carried out in actual laboratory conditions, or when conducting a virtual experiment.
- **Theory**: To view key theoretical concepts governing the experiment.
- **Virtual Experiment**: To conduct a virtual experiment.
- **Data Analysis**: To view a general description of the methods used to analyze the data obtained from the virtual experiment.
- **Discussion**: A list of discussion questions related to the results and data analysis.
- **Online Resources**: Web addresses of manufacturers of industrial equipment related to the experiment
- **References**: A list of suggested references.
- **MyJournal**: To open a Notepad® file to enter any data or to take notes during the experiment (Fig. 1.5), this file may

be saved at the end of the experiment before exiting the program.

- **Exit**: To exit the program.

The **Overview, Industrial Systems, Procedures, Theory, Data Analysis, Discussion, Online Resources** and **References** options include visual and text description of the experiment. For example, in **Industrial Systems**, the user will find the working principle of selected industrial equipment, photographs and animations (Fig. 1.6). In **Data Analysis**, one may find ways to analyze the data (Fig. 1.7).

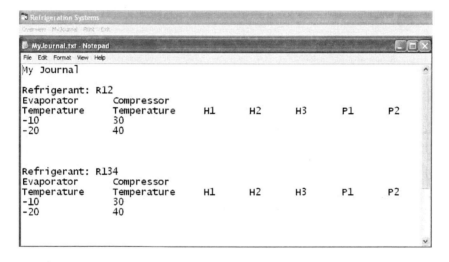

Figure 1.5. The Laboratory **Overview** Screen and the **MyJournal** text file.

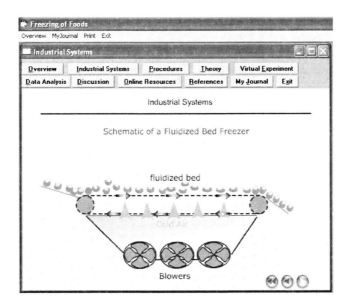

Figure 1.6. An animated view of an industrial fluidized bed freezing system.

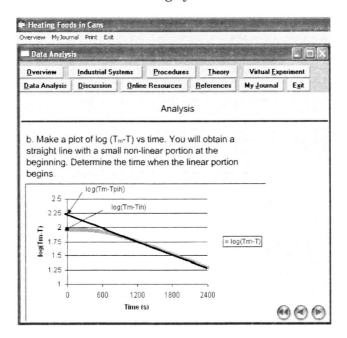

Figure 1.7. Data analysis for determining heat transfer parameters in foods during canning.

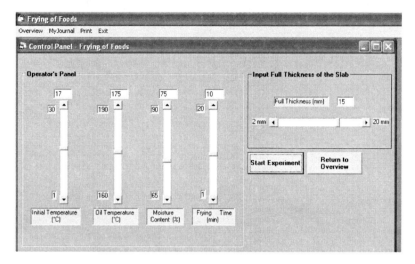

Figure 1.8. Control panel for an experiment to predict
temperature in a food during frying.

1.5 Control Panel

When the **Virtual Experiment** button is clicked, a control panel
screen of the selected experiment will be shown. The control panel
includes the product and process variables that a user may
change or select when a given experiment is run (Fig. 1.8). In
some experiments, the control panel screen is also used to
indicate the results of the experiment conducted (e.g. in Chapter
2, "Pumping Liquid Foods – Energy Requirements of Pumping
Apple Juice"). In other experiments, the control panel screen
opens another screen where some graphical outputs of the
experiment are shown (e.g., in Chapter 15, Food Freezing –
Determining Freezing Time of Potato).

On a control panel, a user generally has the following options
(Fig. 1.9):

- **Start Experiment**: To start the experiment and collect the
  experimental data
- **View Data in Spreadsheet**: To view the data collected
  from the virtual experiment in a spreadsheet and then to
  save it in Microsoft Excel format
- **Return to Overview**: To return to the overview screen.
  The **Overview** option in the menu bar may also be used to
  return to the overview screen

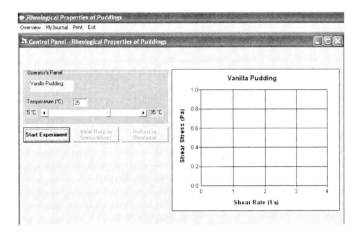

Figure 1.9. Options in the control panel (Start Experiment, View
Data in Spreadsheet, and Return to Overview)

In some laboratory experiments, the **Return to Overview**
button is not seen in the **Control Panel** screen. However, the
follow-up screen, which is seen when the experiment is started,
has these additional options (Fig. 1.10):

- **Stop**: To stop the experiment at any time. Since the data
  collection may be somewhat longer in these experiments
  compared with the others, the user may stop the
  experiment at any time. This option is especially useful
  when a correction of a wrong input of any process or
  product variable is required.
- **View Data in Spreadsheet**: To view the collected
  experimental data in a spreadsheet and then to save it in
  the Microsoft Excel format.
- **Return to Experiment**: To allow the user to return to the
  control panel to rerun an experiment with different input
  conditions.
- **Return to Overview**: To return to the overview screen.
  The **Overview** option in the menu bar may also be used to
  return to the overview screen.

For some of the virtual experiments, when the program is
running, only the **Stop** option is enabled. To enable the use of the
other three options, the experiment must be stopped first.

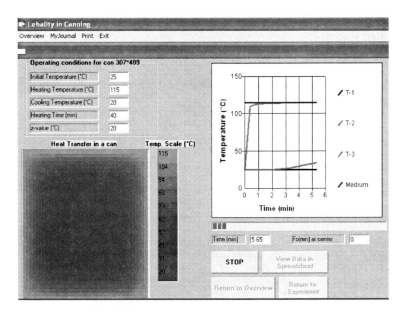

Figure 1.10. Second follow-up screen with additional options (Stop, View Data in Spreadsheet, Return to Experiment and Return to Overview).

When the **View Data in Spreadsheet** option is used, all the experimental data are shown on a spreadsheet (Fig. 1.11). This spreadsheet may be viewed after completing either one experiment, or preferably after running all experiments with different operating conditions. Every time a new experiment is run, the results will be appended to data from a previous experiment.

Note that the spreadsheet showing the results must be converted to Microsoft Excel format before saving using following procedure. To save the spreadsheet, the **Save file as** option from the toolbar (⊞) should be used. When this icon is chosen, the user sees the **Save file as** dialog (Fig. 1.12). It is important that in the **Save as type,** the **Excel 97 (*.xls)** option is selected as shown in Figure 1.12. Any file folder and file name may be chosen to save the experimental data file. In addition to the **Save as** icon, the spreadsheet has also a tool bar with icons for the **Open, Save, Print, Cut, Copy,** and **Paste** options. The user may use them in the same way they are used in a Microsoft Excel spreadsheet. However, keyboard shortcut options (such as **Crtl + X** for cut or **Ctrl + C** for copy, etc.) are NOT included in the features of these spreadsheets. Therefore, use the toolbar icons for these tasks.

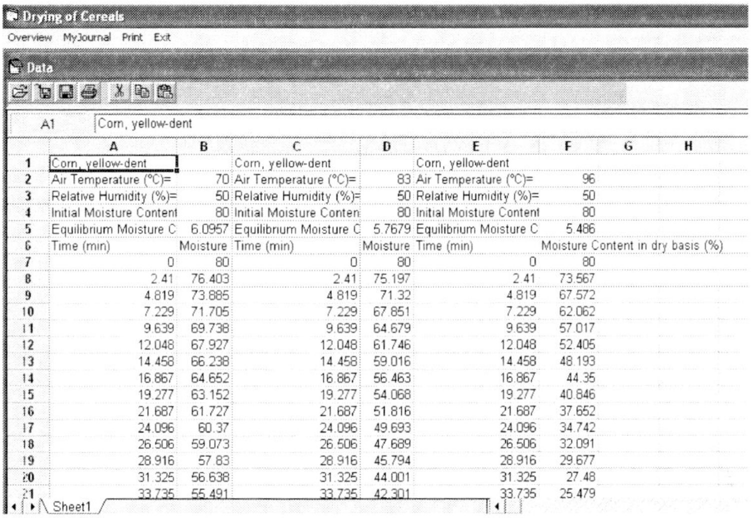

Figure 1.11. Viewing experimental results in a spreadsheet.

Figure 1.12. The **Save file as** dialog to save the spreadsheet in Excel format.

To analyze the saved data with Microsoft Excel, some features require special attention. For example, in the data analysis of some experiments, you may be asked to determine the slope of a linear portion of a curve. The use of the **Trendline** option in Microsoft Excel is the easiest way to accomplish this task.

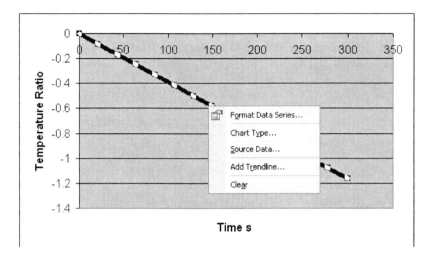

Figure 1.13. Using Trendline analysis of a plot of temperature ratio vs. time.

1.6 Trendline Option in Microsoft Excel

To analyze the data in some experiments, you may need to determine the slope of a linear portion of a curve. To accomplish this, you may use the following steps using a Microsoft Excel spreadsheet:

- To determine the slope of this line, use the **Trendline** feature of Microsoft Excel.
- For the **Trendline** feature, right click mouse when pointing on the line. A list of options will be displayed as shown in Figure 1.13. Select **Add Trendline...** (Fig. 1.13). A window showing the **Type of Trend/Regression** will be shown, select the default option **Linear**. (Fig. 1.14). Next, select the second tab in this window named **Options.** In the next window, select **Display equation** and **Display R-squared value on chart** (Fig. 1.15), continue by selecting **OK** button to display the regression equation on the chart with its R^2 value as shown in Figure (1.16).

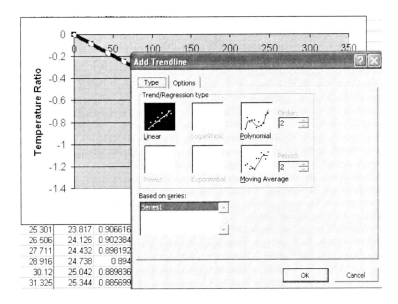

Figure 1.14. The *Trend/Regression type* window of the Trendline feature in Microsoft Excel. *Linear* trend is highlighted as a default option.

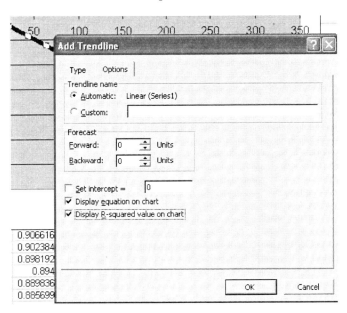

Figure 1.15. The *Options* tab of the Trendline feature in Microsoft Excel, and selection of *Display equation* and *R-squared value on chart.*

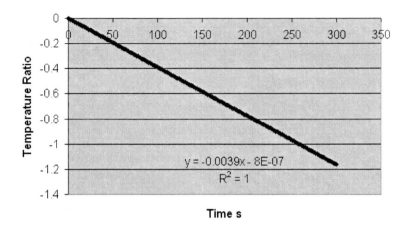

Figure 1.16. Displaying equation and R-squared value on chart for the plot.

- The equation shown on the chart has the form of a straight line equation:

$$y = mx + c$$

where m is the slope and c is the intercept. For example, for the equation shown in Fig. 1.16,

slope, m = – 0.0039

and,   intercept,   c = – 8 × 10⁻⁷

You may use the slope and intercept values for further analysis.

Chapter 2 Pumping Liquid Foods – Energy Requirements of Pumping Apple Juice

Transportation of liquid foods in a commercial food processing plant is an essential operation, as in pumping of milk through various processing equipment in a dairy. For safe and hygienic operation, liquid foods are transported in pipes. A typical piping system consists of cylindrical pipes, a pump, and other elements (also called fittings) such as elbows, tees, and valves.

A pump is a key component of the piping system that provides the necessary energy for the liquid to flow. The energy requirements for pumping are influenced by the physical properties of the liquid, such as viscosity and density, change in elevation, friction losses, and the number of fittings used in the piping system. By reducing the friction losses or optimal selection of fittings, energy requirement for pumping may be reduced (Barbosa-Canovas et al., 1997; Singh and Heldman, 2001).

In a simple liquid transport system between two locations, energy requirements are given by the Bernoulli equation:

$$gz_1 + \frac{P_1}{\rho} + \frac{v_1^2}{2} + W = gz_2 + \frac{P_2}{\rho} + \frac{v_2^2}{2} + F \qquad 2.1$$

where g is acceleration due to gravity (m/s²); P_1 and P_2 are pressure at points 1 and 2 (Pa); z_1 and z_2 are vertical coordinates at locations 1 and 2 (m); v_1 and v_2 are velocities of the liquid food at locations 1 and 2 (m/s); F is energy loss due to friction and fittings used in the system (J/kg); ρ is density of the liquid food (kg/m³); and W is work done by the pump (mechanical energy transferred to the liquid food) (J/kg).

The energy losses are expressed as:

$$F = \frac{2 f v^2 L}{D} + \sum \frac{k_f v^2}{2} \qquad 2.2$$

where, D is diameter of the pipe (m); L is total length of the pipe (m); f is Fanning friction factor; and k_f is unique friction loss coefficient for any particular fitting.

The friction factor, f, is obtained from the following relationships based on whether the flow is laminar or turbulent ($N_{Re} < 2100$ or $N_{Re} > 10000$), respectively.

$$f = \frac{16}{N_{Re}} \qquad N_{Re} < 2100 \qquad 2.3$$

$$\frac{1}{\sqrt{f}} \approx -3.6\log\left[\frac{6.9}{N_{Re}} + \left(\frac{\varepsilon/D}{3.7}\right)^{1.11}\right] \qquad N_{Re} > 2100 \qquad 2.4$$

where, ε/D is relative roughness; N_{Re}: Reynolds Number $= \dfrac{\rho v D}{\mu}$, dimensionless; and μ is dynamic viscosity (kg/ms).

In this laboratory exercise, we will conduct an experiment to determine the energy requirement to pump apple juice in a piping system using a smooth pipe ($\dfrac{\varepsilon}{D} = 0$). The pumping requirement will be determined as the product of the mechanical energy transferred (J/kg) and mass flow rate of the liquid food (kg/s) for a given system.

2.1 Objectives:
1) To calculate the energy requirement of pumping apple juice.
2) To determine the change in energy requirements for pumping with respect to mass flow rate, the total length of the pipes, and number and types of fittings used in the pumping system.

2.2 Materials and Methods:
In a laboratory experiment, we will pump apple juice in a steel pipe (Schedule 40, 2.5 cm diameter) from a storage tank to a heat exchanger (Fig. 2.1). A centrifugal pump is used for pumping. Three 90° standard elbows are used to raise the elevation of the exit discharge to 9 m above the datum.

For the virtual experiment, a suggested list of process variables is given in Table 2.1.

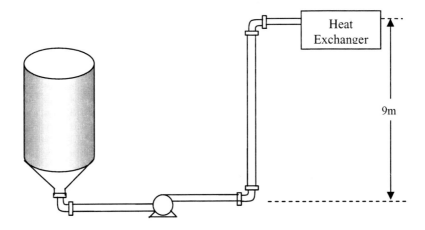

Figure 2.1. Pumping apple juice in a steel pipe system between a storage tank and the heat exchanger.

Table 2.1. Suggested process variables

Trial	Mass Flow Rate (kg/s)	Total Pipe Length (m)
1	1	30
2	5	30
3	10	30
4	1	60
5	5	60
6	10	60
7	1	90
8	5	90
9	10	90

The following steps are used to obtain the experimental data using the virtual experiment.

1) From the introductory screen **Laboratory Overview,** click on appropriate buttons to view **Overview, Industrial Systems, Theory**, and **Procedures** that describe various aspects of the experiment. Then, click on **Virtual Experiment** to conduct the experiment.

2) Select the **Steel Pipe (Schedule 40)** option from the **Control Panel** screen.

3) Select the appropriate check box(es) for the fittings. For our example, select **Elbow, 90°, Standard**. Enter the number of fittings in the respective text boxes (Fig. 2.2).

4) Enter the steel pipe diameter, total elevation for pumping, mass flow rate of juice, and length of pipe. In our example, the pipe diameter is 2.5 cm, and the total elevation is 9 m. The mass flow rate and pipe length are selected from Table 2.1.

2.3 Results:

After entering all the data, click on the **Calculate** button (Fig. 2.2). The results for theoretical pumping requirement will appear in the text box. You will need to record this value for each trial condition of Table 2.1 for further analysis. For this purpose, you may use the **MyJournal** option. To open and use **MyJournal,** point and click on **MyJournal** in the menu bar. Enter the calculated theoretical pumping requirement in the table (Fig. 2.3).

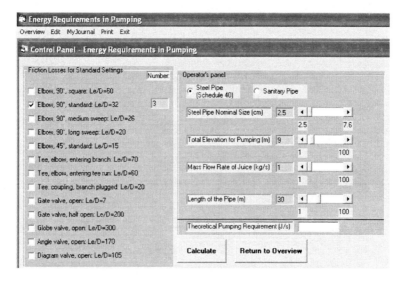

Figure 2.2. Control panel to calculate the theoretical pump power for a piping system.

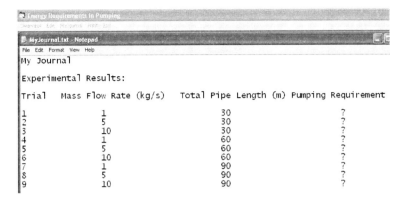

Figure 2.3. Use of **MyJournal** Notepad® file to enter and save experimental results.

Caution: You must save the **MyJournal** file in any folder on your computer's hard disk before exiting the program, otherwise the entered information will be lost.

For additional experiments, you may design different pumping systems with the desired number and types of fittings and determine their effect on the energy requirements.

2.4 Analysis of Results:

In this experiment, you obtained the pumping energy requirement for different conditions of mass flow rates and pipe lengths. The **Data Analysis** option may also be used to view the steps for data analysis. After completing all the trials given in Table 2.1,

1. Plot the change of pumping energy requirement versus total pipe length and mass flow rate of apple juice, and
2. Determine the change of pumping energy requirement with respect to different variables, such as pipe diameter (nominal size) and total pumping elevation.
3. Repeat trials by keeping the mass flow rate and pipe length constant and varying the number and type of pipe fittings.

2.5 Discussion:

1. How does the flow rate affect the pumping energy requirement?
2. What is the effect of total pipe length on pumping energy requirement?
3. What is the influence of fittings on the pumping energy requirements? Which fittings have the most prominent effect?

2.6 Online Links:

Alfa Laval:
> http://www.alfalaval.com

The Pump Center:
> http://www.pumpcentre.com

Fluid Engineering Center:
> http://www.bhrgroup.co.uk

Kraft Unit Operations:
> http://www.kraftunitops.com/unit_operation.html

Jensen Fittings:
> http://www.jensenfittings.com

Balflex-Balfit Stainless Steel Fittings:
> http://www.balflex.com

2.7 Bibliography:

Barbosa-Canovas, G.V., Ma, L., and Braletta, B. 1997. Food Engineering Laboratory Manual. 1997. Technomic Publishing Company, Inc., Lancaster, PA.

Munson, B.R., Young, D.F., and Okiishi, T.H.1998. Fundamentals of Fluid Mechanics. John Wiley and Sons, New York.

Singh, R.P. and Heldman, D.R. 2001. Introduction to Food Engineering. Academic Press. London.

Smits, A.J. 2000. A Physical Introduction to Fluid Mechanics. John   Wiley   and   Sons,   Inc.,   New   York.

Chapter 3 Rheological Properties of Foods – Determining Flow Properties of Vanilla Pudding

The rheological properties of a food product are important in describing its flow and deformation behavior. While liquid foods such as milk and fruit juices show simple flow characteristics, more viscous materials such as ketchup, tomato paste, and mayonnaise flow in a more complicated way (Singh and Heldman, 2001). According to the flow behavior, the fluid foods may be classified into two general groups: Newtonian and non-Newtonian fluids. For Newtonian fluids, the shear stress is a linear function of shear rate (Singh and Heldman, 2001):

$$\sigma = -\mu \frac{dv}{dy} \qquad\qquad 3.1$$

where σ is the shear stress (Pa), μ is the viscosity (Pa-s), and $\dfrac{dv}{dy}$ is the shear rate (1/s).

Viscosity is a property related to the resistance to flow. For example, the viscosity of honey is higher than that of milk. Viscosity may also be defined as a property describing the magnitude of resistance due to shear forces within the fluid.

In Newtonian fluids, the relationship between the shear stress and shear rate is linear. In non-Newtonian fluids, this relationship is nonlinear. The non-Newtonian fluids are classified as dilatant (shear-thickening), pseudoplastic (shear-thinning), Bingham plastic, and Casson-type plastic fluids. A general mathematical expression that describes the characteristics of fluid rheology is the **Herschel-Bulkley** model (Singh and Heldman, 2001):

$$\sigma = \sigma_0 + K \left(\frac{dv}{dy} \right)^n \qquad\qquad 3.2$$

where K is the consistency coefficient (Pa-s^n), and n is the flow behavior index with a value between 0 and 1. For $n = 1$ and yield stress or $\sigma_0 = 0$, the flow characteristics are those of a Newtonian fluid given in Eq. 3.1 (Saravacos and Kostrapoulos, 1996).

The rheological properties of a fluid are measured using a rheometer. A common configuration of this instrument is a

Table 3.1. Suggested conditions for the experiment

Experiment Number	Temperature (°C)
1	10
2	20
3	30

coaxial cylindrical rheometer. In this rheometer, a solid cylinder rotates in a fluid kept inside a stationary cup, and the shear stress is recorded vs. shear rate.

In this laboratory exercise, we will conduct an experiment with a coaxial cylindrical rheometer using vanilla pudding at different temperatures. The shear stress values will be experimentally measured at different shear rates. Data analysis will include the calculation of the consistency coefficient and flow behavior index of the given samples at different temperatures.

3.1 Objectives:
1. To determine the consistency coefficient and flow behavior index values of vanilla pudding.
2. To determine and describe the effect of temperature on the rheological properties of vanilla pudding.

3.2 Materials and Methods:
In a laboratory experiment, we deposit a sample of vanilla pudding in the cylindrical cup of a coaxial cylinder rheometer. The temperature of the pudding is maintained by pumping water at a constant temperature in a jacket surrounding the sample cup. The rheometer is programmed to automatically lower the rotor in the pudding sample. The shear stress at different shear rates is recorded.

For a virtual experiment, you will use three temperatures to determine rheological properties of vanilla pudding, as shown in Table 3.1.

The following steps are used to obtain the experimental data using the virtual experiment.
1. From the introductory screen **Laboratory Overview,** click on appropriate buttons to view **Overview, Industrial Systems, Theory, and Procedures** that describe various aspects of the experiment. Then, click on **Virtual Experiment** to conduct the experiment.
2. After selecting the desired temperature by using horizontal scroll bars (or directly entering values in the text box), click on the **Start Experiment** button to begin the experiment.

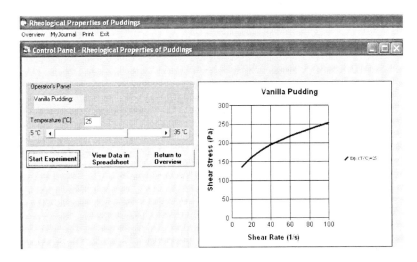

Figure 3.1. Control panel to select experimental variables.

The shear stress vs. shear rate data will be shown as a plot (Fig 3.1). Repeat the experiment for different temperatures.

3.3 Results:
Once the experiment for the selected conditions in the previous screen is completed for the given temperatures, you will need to save the data in Excel format for further analysis. To save the data, click on the **View Data in Spreadsheet** button. The data will be shown on a spreadsheet. Save the data in Excel (*.xls) format, as described in Chapter 1.

3.4 Analysis of Results:
In this experiment, you obtained data on shear stress vs. shear rate at different temperatures. To analyze the data:
1. Using the **XY Scatter plot** option in Excel, plot shear stress versus shear rate at different temperatures and note any differences in plots.
2. Use the following steps to determine the consistency coefficient and the flow behavior index (Fig. 3.2):
3.   Eq. 3.2 may be modified as:

$$\ln(\sigma - \sigma_0) = \ln(K) + n\ln(\frac{dv}{dy})$$ 3.3

4. The value of σ_0 must be known to use Eq. 3.3. One procedure is to manually extend the shear stress-shear rate curve until it intersects the y-axis. Another method is to use the following equation to calculate σ_0. This

equation has been specifically derived for vanilla pudding for the conditions of this experiment.

$$\sigma_0 = 0.0001713 \times e^{\left(\frac{E_a}{RT}\right)}$$

3.4

where E_a = 30430 kJ/kg-mol; R = 8.3143 kJ/kg-mol K; T is in Kelvin.

5. Create a plot of $\ln(\sigma - \sigma_0)$ values vs. $\ln\left(\dfrac{dv}{dy}\right)$ for each temperature.
6. Determine the slope and intercept of this plot using the **Trendline** option in Excel.
7. The slope gives the value of the behavior index, n.
8. The exponent of the intercept gives the consistency coefficient (K).

3.5 Discussion:
1. Discuss how the K and n-values change with temperature for vanilla pudding.
2. Assuming that K and n change with respect to temperature following an Arrhenius type equation, calculate the values of Activation Energy (E_{aK} and E_{an}) and coefficients (K_0 and n_0):

$$K = K_0 \exp\left(\frac{E_{aK}}{RT}\right)$$

3.4

$$n = n_0 \exp\left(\frac{E_{an}}{RT}\right)$$

3.5

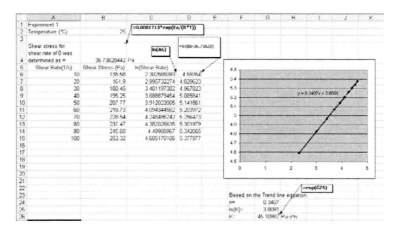

Figure 3.2 Spreadsheet showing calculations to obtain rheological properties.

where K_0 and n_0 are the pre-exponential coefficients; R is the universal gas constant; and T is the temperature in Kelvin.

3.6 Online Links:

Brookfield:
http://www.brookfieldengineering.com
Porpoise:
http://www.porpoise.co.uk
The Society of Rheology:
http://www.rheology.org
Rheology Technical Center:
http://www.rheology-online.com

3.7 Bibliography:

Dodge, D.W. and Metzner, A.B. 1959. Turbulent flow of non Newtonian systems. AICHE J. 5(7), 189-204.

Doublier, J.L. and Lefebvre, J. 1989. Flow properties of fluid food materials. In "Food Properties and Computer-Aided Engineering of Food Processing Systems" (R. P. Singh and A. G. Medina, eds), pp. 245-269. Kluwer Publishers, Dordecht, The Netherlands.

Saravacos, G.D. and Kostaropoulos, A.E. 1996. Engineering properties in food properties simulation. Computers and Chemical Engineering. 20: S461-S466.

Singh R.P. and Heldman, D.R. (2001). "Introduction to Food Engineering," 3rd ed., Academic Press, London.

Steffe, J.F. (1996). "Rheological Methods in Food Process Engineering," 2nd ed., Freeman Press, E. Lansing.

Chapter 4 Temperature Sensors – Response Time of Thermocouples

An important question in selecting a temperature sensor for many industrial applications is: How fast will a temperature sensor respond to a change in the ambient temperature? To answer this question, we need to know the dynamic response characteristics of the sensor or its time constant. The time constant gives information on how fast or slow a given sensor responds to a change in the input signal. If a sensor responds too slowly, there will be a considerable lag in measuring changes in the ambient temperature. This lag in response may be a major limitation of such a sensor, especially when the temperature of a system changes very rapidly such as in sterilizing liquid foods. A sensor with slow response may be adequate when the environmental temperature changes slowly and only an average temperature is required. Nevertheless, the terms "rapid" and "slow" are subjective terms, and they are not very useful in selecting the sensors. In this laboratory exercise, we will determine a quantitative measure of the temperature response, or time constant, of different sensors in different surrounding media.

4.1 Objectives:
1. To determine the response time of different temperature sensors when they respond to a sudden temperature change in the ambient environment.
2. To calculate the time constant of temperature sensors using experimentally obtained data on temperature vs. time.

4.2 Materials and Methods:
In a laboratory experiment, each sensor is first allowed to equilibrate to room temperature ($\approx$20-25°C). Then it is placed in either an ice bath (0°C) or in boiling water ($\approx$100°C) to reach equilibrium again. The temperature vs. time data for each sensor is recorded to determine the time constant values.

In a virtual experiment, you will use four different temperature sensors, A, B, C, and D. Their dynamic responses will be determined in two different surrounding media (boiling water and ice bath). The initial temperature in each experiment will be 20°C. Table 4.1 lists the suggested temperatures and equilibrium conditions for the experiment.

Table 4.1. Suggested conditions for the experiment.

Sensors	Surrounding Medium	Medium Temperature (°C)
Sensor A	Ice bath	0
	Boiling water	100
Sensor B	Ice bath	0
	Boiling water	100
Sensor C	Ice bath	0
	Boiling water	100
Sensor D	Ice bath	0
	Boiling water	100

The following steps are used to obtain the experimental data using the virtual experiment.

1. From the introductory screen **Laboratory Overview,** click on appropriate buttons to view **Overview, Industrial Systems, Theory, and Procedures** that describe various aspects of the experiment. Then, click on **Virtual Experiment** to conduct the experiment.

2. In the control panel (Fig. 4.1), select one or more temperature sensors and enter the initial temperature and medium temperature in the appropriate text box.

3. Select **Start Experiment** to begin the experiment.

4. A plot of temperature vs. time will be shown. Repeat the experiment with each of the other sensors.

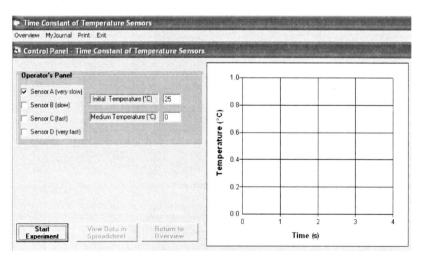

Figure 4.1. Control panel to determine time constant of different temperature sensors.

Time Constant of Temperature Sensors

Overview   MyJournal   Print   Exit

Data1

F13

	A	B	C	D
1	Sensor A		Sensor B	
2	Initial Temperature (°C)=	25	Initial Temperature (°C)=	25
3	Medium Temperature (°C)=	0	Medium Temperature (°C)=	0
4	Time (s)	T (°C)	Time (s)	T (°C)
5	0	25	0	25
6	0.241	24.602	0.241	24.405
7	0.482	24.21	0.482	23.824
8	0.723	23.824	0.723	23.257
9	0.964	23.444	0.964	22.703
10	1.205	23.07	1.205	22.162
11	1.446	22.703	1.446	21.635
12	1.687	22.341	1.687	21.12
13	1.928	21.985	1.928	20.617
14	2.169	21.635	2.169	20.126
15	2.41	21.29	2.41	19.647
16	2.651	20.951	2.651	19.179
17	2.892	20.617	2.892	18.722

Figure 4.2. Experimental data obtained for different sensors.

4.3 Results:
After the **Start Experiment** button is selected, a plot (Fig. 4.1) will show the temperature change for the selected sensor in the given medium.

After completing all the trials given in Table 4.1, view the numerical data by selecting the **View Data in Spreadsheet** button (Fig. 4.2). Save these data in Excel format (*.xls), as described in Chapter 1.

4.4 Analysis of Results:
The response time of a sensor is obtained by determining the time constant. A time constant of a sensor is defined as the time required by a sensor to reach 63.2% of a step change in temperature under a given set of conditions. The time constant is expressed in units of time (e.g., seconds), and it is a property of a given sensor. Its determination is based on the observation that a given sensor will respond in an exponential manner to a sudden change in input (when the surrounding temperature changes suddenly). Mathematically,

$$T = T_m - \left(T_m - T_0\right)e^{\left(-\frac{t}{\tau}\right)}$$

4.1

where:

T is sensor temperature (°C), T_m is surrounding medium temperature (°C), T_0 is initial temperature (°C), t is time (s), and τ is time constant (s).

Rearranging the preceding equation and taking the natural logarithm of both sides results in:

$$\ln\left(\frac{T-T_m}{T_0-T_m}\right) = -\frac{t}{\tau} \qquad 4.2$$

Then, the time constant (τ) is obtained from the slope of a linear plot of $\ln\left(\frac{T-T_m}{T_0-T_m}\right)$ versus t.

In this experiment, you obtained time-temperature data for four temperature sensors subjected to different medium temperatures. To determine the time constant, use the following steps:

1. Determine $\ln\left(\frac{T-T_m}{T_0-T_m}\right)$ at each time in an Excel spreadsheet.

2. Create a plot of $\ln\left(\frac{T-T_m}{T_0-T_m}\right)$ vs. time for each sensor.

3. Determine the slope of each line using the **Trendline** option in Excel.

4. Calculate the time constant for each sensor from the following expression:

$$\tau = -\frac{1}{slope}$$

4.5 Discussion:

1. Discuss any observed differences in the time constant values found for different medium temperatures.
2. Which characteristics of the sensors have the most influence on the time constant?
3. If a sensor has a time constant of 3 seconds, how long would it take it to respond to 99% of a sudden change in ambient temperature?

4.6 Online Links:

Temperature Sensors:
    http://www.temperatures.com/
Temperature Sensors and Assemblies:

http://www.minco.com/sensors.php
Thermometrics Corporation:
    http://www.thermometricscorp.com/
Honeywell Control Products:
    http://content.honeywell.com/sensing/prodinfo/temperat
ure/
Industrial Temperature Sensors:
    http://www.its.irl.com/
Temperature Sensors:
    http://www.temperatureworld.com/shci.htm
Dwyer Temperature Sensors:
    http://www.dwyer-
inst.com/htdocs/temperature/TemperatureSensors.html
Process Measurement and Control:
    http://www.omega.com

4.7 Bibliography:

Holman, J.P. 2001. Heat Transfer, 9th edition. McGraw Hill, Inc., New York.

Doeblin, E.O. 1983. Measurement Systems, Application and Design. McGraw Hill Book Company. New York.

Chapter 5 Convective Heat Transfer – Determining Heat Transfer Coefficient in Air and Water

The thermophysical properties of foods are important in modeling industrial food processing operations such as heating, cooling, refrigeration, frying, thawing, and freezing, (Erdogdu et al., 1998). The properties used to model heat transfer in these processes are thermal conductivity (k), specific heat (c_p), density (ρ), and thermal diffusivity (α). These properties are influenced by the composition of a food material. In addition to the properties of food, convective heat transfer coefficient (h-value) of the heating or cooling fluid is an important parameter that affects the heating and cooling rates. The h-value depends on the thermophysical properties of the fluid surrounding the food, characteristics of the food (shape, dimensions, surface temperature, and surface roughness), and the characteristics of the fluid flow (velocity, viscosity and turbulence) (Rahman, 1995).

A wide range of h-values for different processes have been reported in the literature (Nicolai and DeBaerdemaeker, 1996): 300 W/m² K for hydrocooling of fruits and vegetables, 6 to 68 W/m² K for forced air cooling, 256 to 10,000 W/m² K for sterilization in steam retorts, 500 to 700 W/m² K for immersion cooling with water, 14 to 68 W/m²K for heating in home and rotary reel ovens, and 44 to 170 W/m² K for heating in commercial air-impingement ovens.

The h-value is often calculated with empirical correlations between the dimensionless numbers (Nusselt number: N_{Nu}; Reynolds number: N_{Re}; Prandtl Number: N_{Pr}; Grashof Number: N_{Gr}).

$$N_{Nu} = f\left(N_{\mathrm{Re}}, N_{\mathrm{Pr}}\right) \qquad 5.1$$

$$N_{Nu} = f\left(N_{Gr}, N_{\mathrm{Pr}}\right) \qquad 5.2$$

Eqs. (5.1) and (5.2) are for forced or free convective heat transfer, respectively. Numerous expressions exist in the literature for the calculation of h-value for regularly shaped objects (Cengel, 1998; Geankoplis, 1993; Incropera and Dewitt, 1990). However, specific expressions do not exist for the calculation for irregular geometries. Many times, irregular shapes may be approximated by regular shapes enabling the use of the existing correlations. However, a preferred method is to determine the h-value by experiments.

In this laboratory exercise, we will determine the h-value for spherical shaped objects when submerged in hot water, stationary air, or forced air at different temperatures using the quasi-steady-state method. The procedure involves solid aluminum spheres of different diameters that are heated or cooled in a given environment, and their temperature change vs. time is recorded. Data analysis includes the calculation of h-value from the experimental data for different conditions. The following description provides the basis of this method.

During heating or cooling of a solid object, the temperature gradient inside a solid may be considered negligible if there is a low Biot number ($N_{Bi} < 0.1$). In such a case, the ratio of the internal resistance to heat transfer in the solid to the external resistance to heat transfer in the fluid is given by Eq. 5.3:

$$N_{Bi} = \frac{hd_c}{k} < 0.1 \qquad\qquad 5.3$$

where h is the convective heat transfer coefficient, W/m²-K; d_c is the characteristic length (radius for a sphere), m; and k is thermal conductivity, W/m-K.

First, we will write an energy balance around a solid object made of a high thermal conductivity material (e.g., aluminum). It is assumed that the object is suddenly immersed into a fluid of different temperature. The energy balance equation gives:

$$hA(T_m - T) = mc_p \frac{dT}{dt} \qquad\qquad 5.4$$

where A is surface area (m²), m is mass (kg), c_p is heat capacity heat of the material, T_m is surrounding fluid temperature (°C), T is temperature of the material (°C), and t is the time (s).

The solution of Eqn. (5.4) is:

$$\ln\left[\frac{T - T_m}{T_i - T_m}\right] = -\frac{hA}{mc_p} t \qquad\qquad 5.5$$

by using the initial condition:

$$T(0) = T_i \qquad\qquad 5.6$$

where T_i is the initial temperature (°C).

The heat transfer coefficient may be calculated from the slope of the line $\ln[(T-T_m)/(T_i - T_m)]$ vs. time (t). This method may be used with any shape when the object's mass, surface area, heat capacity, and the experimental time-temperature data are known.

$$h = -\frac{(slope)mc_p}{A}$$ 5.7

5.1 Objectives:

1. To measure the change in temperature of an aluminum sphere when it is heated or cooled in either water or air.
2. To determine the heat transfer coefficient for a spherical object heated or cooled in either hot water or cold air by using a quasi-steady-state method.
3. To examine the effect of product size, water temperature, air temperature, and air velocity on the convective heat transfer coefficient.

5.2 Materials and Methods:

In a laboratory experiment, we use aluminum spheres of different radii embedded with thermocouples. Aluminum has a very high thermal conductivity (k = 204 W/m K), therefore the assumptions for the quasi-steady-state method will hold. The temperature change of the aluminum spheres is recorded using a data acquisition system.

In a laboratory experiment, the spherical transducer is first equilibrated at a given initial temperature. It is then submerged in a fluid and the change in temperature is recorded.

For the virtual experiment, we have selected spherical transducers of three different diameters, three air velocities and three different hot water temperatures. Table 5.1 lists the process variables to be used in the experiment.

The experiments consist of the following parts:

1. Determine the h-value for hot water (98°C), still air (20°C), and forced air (2 m/s, 20 °C) using a sphere of 15 mm in radius (initial temperature of the aluminum sphere set at 20 °C for heating and 95 °C for cooling).

Table 5.1. Suggested process variables.

Spherical Transducer Radius (mm)	Hot water temperature (°C)	Air Velocity
5	90	1
10	95	5
20	98	10

2. To investigate the effect of product size, spheres of 5, 10, and 20 mm in radii are heated in hot water (90°C), and the h-values are determined.
3. To examine the effects of air velocity on heat transfer coefficient, spheres of 20 mm in radius heated to a uniform internal temperature of 95°C are used. These spheres are cooled with air at 20°C at different velocities (1, 5 and 20 m/s). (See Table 5.2 for suggested conditions).

The following steps are used to obtain the experimental data using the virtual experiment.

1. From the introductory screen **Laboratory Overview,** click on appropriate buttons to view **Overview, Industrial Systems, Theory, and Procedures** that describe various aspects of the experiment. Then, click on **Virtual Experiment** to conduct the experiment.

Table 5.2. Suggested conditions for the process and product variables.

Trial	Expt No	Radius - Initial Temperature	Still Air	Forced Air	Hot Water	h-val.
1st	1	15 mm – 20°C	-	-	98°C	
	2	15 mm – 95°C	20°C	-	-	
	3	15 mm – 95°C	-	2 m/s – 20°C	-	
2nd	1	5 mm – 20°C				
	2	10 mm – 20°C	-	-	98°C	
	3	20 mm – 20°C				
3rd	1			1 m/s – 20°C		
	2	20 mm – 95°C	-	5 m/s – 20°C	-	
	3			20 m/s – 20°C		

2. From the **Control Panel**, select the heating/cooling medium of interest. Enter the radius of the aluminum sphere, initial temperature of the sphere, and medium temperature (Fig. 5.1).
3. Select **Start Experiment** button to begin the experiment.

5.3 Results:

After the **Start Experiment** button is selected, the change of temperature vs. time for the aluminum sphere will be shown on a plot. Repeat the trials using conditions earlier given in Table 5.2. The numerical data may be viewed by selecting the **View Data in Spreadsheet** button.

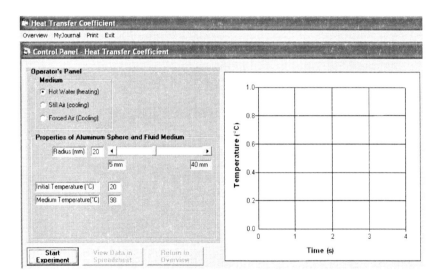

Figure 5.1. Control panel for the virtual experiment of the convective heat transfer coefficient determination.

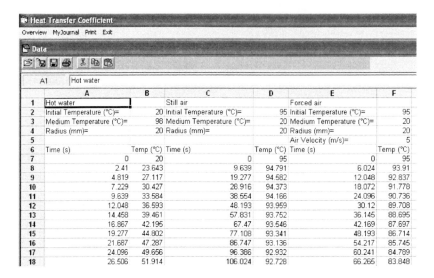

Figure 5.2. A general view of time-temperature data in a spreadsheet for different experimental conditions.

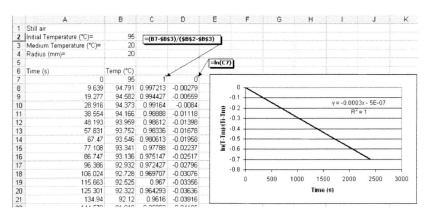

Figure 5.3. Analysis of data to determine the slope of the temperature ratio vs time plot.

5.4 Analysis of Results:

In this experiment, you obtained temperature-time data for different conditions (Fig. 5.2) and saved them in Excel format (*.xls). Use the following steps to determine the h-value using quasi-steady state method. You may also use the **Data Analysis** button to view steps for analyzing the data.

The data analysis involves the following steps.

1. Using the temperature data, calculate the ratio of $\ln\left[\dfrac{T-T_m}{T_i-T_m}\right]$ in a separate column.

2. Create a scatter plot of $\ln\left[\dfrac{T-T_m}{T_i-T_m}\right]$ versus time.

3. In step 2, you will obtain a straight line. If there is some nonlinear portion toward the end of the heating/cooling period, you should delete that part of the data and redraw using only the linear part, as shown in Fig. 5.3.

4. Using the **Trendline** feature in Excel, determine the slope of the linear line (Fig. 5.3).

5. Determine the h-value using the calculated slope value, total surface, total mass and heat capacity of the sphere using Eq. 5.7.

5.5 Discussion:
1. Discuss how the sphere size influences the convective heat transfer coefficient.
2. Discuss the effect of air velocity on the convective heat transfer coefficient.
3. List any limitations in the experimental procedures that may be sources of error?

5.6 Online Links:
Correlations for convective heat transfer:
http://www.cheresources.com/convection.shtml
Evaluation of different heat transfer coefficient definitions:
http://www.electronics-cooling.com/Resources/EC_Articles/JUN95/jun95_04.htm
Heat transfer in a sphere:
http://www.femlab.com/showroom/interactive/chem/sphere/index.php
Heat transfer example:
http://www.uts.com/products/compheat.html

5.7 Bibliography:
Cengel, Y.A. 1998. Heat Transfer: A practical Approach. McGraw-Hill, Inc. Highstown, N.J.

Erdogdu, F., Balaban, M.O., and Chau, K.V. 1998. Automation of heat transfer coefficient determination: Development of a Windows-based software tool. Food Technology in Turkey. 10: 66-75.

Geankoplis, C.J. 1993. Transport Processes and Unit Operations. Prentice Hall, Inc. Englewood Cliffs, New Jersey.

Incropera, F.P. and Dewitt, D.P. 1990. Fundamentals of Heat and Mass Transfer. John Wiley and Sons, Inc. New York, N.Y.

Nicolai, B.M. and DeBaerdemaeker, J. 1996. Sensitivity analysis with respect to the surface heat transfer coefficient as applied to thermal process calculations. Journal of Food Engineering. 28: 21-33.

Rahman, S. 1995. Food Properties Handbook. CRC Press Inc. Boca Raton, FL, U.S.A.

Chapter 6 Heat Exchangers – Heating Milk in Tubular Heat Exchangers

Heat exchangers are commonly used in the food industry to heat or cool foods. Heat exchangers are classified into two general groups:

- Contact type (steam infusion and steam injection), and
- Noncontact type (scraped surface, shell and tube, tubular and plate).

In noncontact type heat exchangers, the product being heated or cooled and the heating or cooling medium are kept separated by a thin wall. On the other hand, in contact-type heat exchangers, there is a direct contact between the product and the heating or cooling medium (Singh and Heldman, 2001). Of these, the simplest noncontact type heat exchanger is a double pipe heat exchanger consisting of two concentric pipes where the product flows in the inner pipe, and the heating or cooling medium is pumped through the annular space.

In designing heat exchangers, prediction of the outlet temperature of the two streams exiting a heat exchanger is an important task. Steady-state heat transfer in a heat exchanger is assumed. There are two common methods used to analyze heat transfer in a heat exchanger: log mean temperature difference (LMTD) and effectiveness-NTU (number of thermal units) (Cengel, 1998). The temperature difference between the two fluid streams varies along the length of the heat exchanger. This difference is expressed as a log mean temperature difference:

$$\Delta T_{LMTD} = \frac{\Delta T_1 - \Delta T_2}{\ln\left(\dfrac{\Delta T_1}{\Delta T_2}\right)} \qquad 6.1$$

where T_1 and T_2 are the temperature differences between the two fluids at the inlet and outlet of the heat exchanger.

The LMTD method is easy to use as long as the inlet and outlet temperatures of the two fluid streams are known, or they may be determined using an energy balance between the two fluids:

$$Q = m_h c_{p_h}\left(T_{h,in} - T_{h,out}\right) = m_c c_{p_c}\left(T_{c,out} - T_{c,in}\right) = UA\Delta T_{lm} \qquad 6.2$$

where subscripts "h" and "c" represent the hot and cold fluids, respectively; U is overall heat transfer coefficient (W/m²-K); c_p is the specific heat capacity (kJ/kg°C); m is mass flow rate (kg/s);

and A is the overall surface area of the heat exchanger based on the inner tube (m^2).

The effectiveness-NTU method is used to determine the outlet temperatures of the two streams for known mass flow rates (kg/s) and inlet temperatures for a given heat exchanger. This method is based on heat transfer effectiveness, ε, given as:

$$\varepsilon = \frac{Q}{Q_{max}} = \frac{Actual\ heat\ transfer\ rate}{Maximum\ possible\ heat\ transfer\ rate} \qquad 6.3$$

where ε is effectiveness value. The actual heat transfer rate is given by Eq. 6.2, and maximum possible heat transfer rate is given by Eq. 6.4:

$$Q_{max} = C_{min}\left(T_{h_{in}} - T_{c_{in}}\right) \qquad 6.4$$

where $C = m \times c_p$ is the heat capacity (kJ/°C). The reason C_{min} is used in Eq. 6.4 is that a fluid with lower heat capacity will experience a greater temperature change. Equations describing effectiveness of different types of heat exchangers are available in Cengel (1998).

In this laboratory exercise, we will conduct an experiment to determine the influence of flow rates on the outlet temperatures of a heating medium (hot water) and a liquid food (cold milk) using a double pipe heat exchanger.

6.1 Objectives:

1) To determine the outlet temperatures of milk and hot water exiting from a tubular heat exchanger.
2) To examine the role of parallel and counterflow options in operating a double pipe heat exchanger and their effectiveness.
3) To determine the overall heat transfer coefficient for a double pipe heat exchanger.

6.2 Materials and Methods:

In this experiment, we will consider heating cold milk using hot water in a double pipe heat exchanger with parallel and counterflow options.

In a laboratory experiment, we use a tubular heat exchanger, with embedded thermocouples to measure temperatures of inlet and outlet streams. The flow rates of product and heating/cooling medium are measured.

For the virtual experiment, we will consider a double pipe heat exchanger with counter and parallel flow options. The suggested

Table 6. 1. Suggested process variables.

Mass Flow rate of Milk in the inner tube (kg/s)	Mass Flow Rate in the outer tube for Parallel Flow Option (kg/s)	Mass Flow Rate in the outer tube for Counter- Flow Option (kg/s)
	Hot Water	Hot Water
0.5	0.5	0.5
	1	1
	2	2
	4	4
1	1	1
	2	2
	4	4
	8	8

inlet temperatures are 90°C for the hot water stream and 4°C for the cold milk. The outer and inner pipe radii are taken as 3 and 2 cm, while the total length of the heat exchanger is 10 m. Table 6.1 lists the suggested variables for each trial. The following steps are used to obtain the experimental data using the virtual experiment.

1) From the introductory screen **Laboratory Overview,** click on appropriate buttons to view **Overview, Industrial Systems, Theory, and Procedures** that describe various aspects of the experiment. Then, click on **Virtual Experiment** to conduct the experiment.

2) In the **Control panel** (Fig. 6.1), enter the inlet temperatures and mass flow rates of the product (milk) and heating fluid (hot water) streams, type of flow, inside pipe radius, and length of the heat exchanger.

3) Click on **Calculate** to determine the outlet temperatures of both fluid streams.

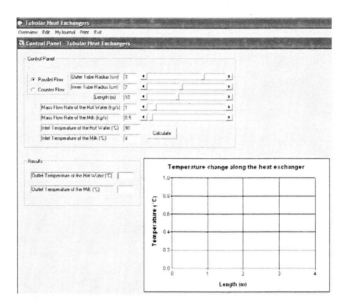

Figure 6.1. Control panel to calculate the outlet temperatures of milk and hot water streams in a double pipe heat exchanger.

6.3 Results:

Upon clicking on the **Calculate** button, the outlet temperatures are shown in the text boxes for hot and cold fluid streams. The graph shows the change in temperature of the hot and cold fluid streams along the length of the heat exchanger. Record the exit temperatures for each trial in Table 6.1. To use **MyJournal** for this purpose, highlight the exit temperature and copy using Ctrl+C, then click on **MyJournal** in the menu bar. Paste (Ctrl+V) the copied data in respective locations in the table.

   Caution: You must save the **MyJournal** file in any folder on your computer's hard disk before exiting the program, otherwise the entered information will be lost.

6.4 Analysis of Results:

Using the results saved in **MyJournal**, calculate the following,
- the logarithmic mean temperature difference using Eq. 6.1,
- the overall heat transfer coefficient (U value) using Eqs. 6.1 and 6.2
- the effectiveness values using Eqs. 6.2 to 6.4 with the calculated outlet and inlet temperatures for each trial for counter- and parallel flow conditions.
- Compare counterflow and parallel flow heat exchangers based on the logarithmic mean temperature difference, overall heat transfer coefficient, and effectiveness values.

6.5 Discussion:
1. Based on the calculated values of the logarithmic mean temperature difference, overall heat transfer coefficient and effectiveness, which flow option would you choose for a double pipe heat exchanger? Why?
2. Are there are any situations when a parallel flow option may be preferred instead of a counterflow heat exchanger?

6.6 Online Links:
Inproheat Industries Ltd.:
   http://www.inproheat.com/heat_ex.htm
AB&CO Heat Exhangers:
   http://www.abco.dk/range.htm
Yula Corporation:
   http://www.yulacorp.com/industrial.html
Alfa Laval:
   http://www.alfalaval.com
Industrial Quick Search Inc.:
   http://www.heatexchangers.org/
Specialized Mechanical Equipment:
   http://www.specialized-mechanical.com/products.html

6.7 Bibliography:
Cengel, Y.A. 1998. Heat Transfer: A Practical Approach. McGraw-Hill, Inc. Highstown, N.J.

Holman, J.P., 2001. Heat Transfer, 9th edition. McGraw Hill, Inc., New York.

Singh, R.P. and Heldman, D.R. 2001. Introduction to Food Engineering. Academic Press. London.

Chapter 7 Heating Liquid Foods – Heating Tomato Juice in a Steam-Jacketed Kettle

During heating or cooling of foods, we are often interested in knowing how the temperature of a product changes with time. The heating or cooling rates vary with different food materials and conditions used to accomplish heat transfer. In this laboratory, we will determine the change in temperature when heating tomato juice in a steam-jacketed kettle. This exercise will help us in identifying some of the key equipment and process variables that influence the heating of liquid foods in kettles. The heat transfer into a food product during the early stages of heating is called transient (or unsteady-state) heat transfer. In transient heat transfer, the temperature varies with both location and time in contrast to steady-state heat transfer, where the temperature variation is only with location. In analyzing problems involving transient heat transfer, knowing the relative importance of heat transfer at the surface of a product undergoing heating (or cooling) in comparison to the heat transfer in the interior of a product is an important step. A quantitative description of this comparison is obtained by determining the Biot Number (N_{Bi}):

$$N_{Bi} = \frac{hd_c}{k} \qquad 7.1$$

where h is convective heat transfer coefficient, d_c is characteristic dimension (e.g., radius for a sphere and an infinite cylinder, half-thickness for an infinite slab), and k is thermal conductivity of the product.

When the Biot Number is smaller than 0.1, there is a negligible resistance of heat transfer in the interior of the product in comparison to that at the outside surface. This condition is also referred to as **lumped system** and the following expression may be obtained to describe temperature change with time by conducting a simple energy balance:

$$\frac{T - T_m}{T_i - T_m} = e^{-\frac{hA}{\rho V c_p} t} \qquad 7.2$$

where T_m is heating or cooling medium temperature (°C); T_i is initial temperature of the product (°C); A is total surface area (m²); ρ is the density (kg/m³); h is the convective heat transfer coefficient (W/m²°C); and c_p is the heat capacity of the product (kJ/kg°C) (Singh and Heldman, 2001).

Lumped system analysis is valid only when the temperature distribution within the food is nearly uniform, and it changes with time but not with location. This condition is normally encountered in materials with high thermal conductivity. Another way to obtain such a condition is when a liquid food in a container is well-stirred (Singh and Heldman, 2001).

In this exercise, we will conduct an experiment by heating tomato juice in a well-stirred, hemispherical steam-jacketed kettle as an example of the lumped system analysis. The temperature change of the tomato juice will be obtained for different processing conditions, and the effect of different processing variables on the heating rate will be determined. Heating rate is described by the slope of the $\ln\left[\dfrac{T-T_m}{T_i-T_m}\right]$ vs. time (t) curve.

7.1 Objectives:

1. To determine the temperature change of tomato juice heated in a well-stirred hemispherical steam jacketed kettle.
2. To determine the effect of kettle surface temperature (steam temperature) and the convective heat transfer coefficient on the temperature change of tomato juice during heating.

7.2 Materials and Methods:

In this experiment, we will heat tomato juice in a steam-jacketed, hemispherical kettle. The initial temperature of the tomato juice will be kept constant.

Typical controls for this kind of heating process include the steam pressure or temperature (which is usually the same as the kettle surface temperature) and heating time. For this experiment, the kettle radius and the heating time are kept constant at 0.5 m and 10 min, respectively. The agitation of juice affects the convective heat transfer coefficient on the inside surface of the kettle. Thus, the value of heat transfer coefficient is an indirect measure of the amount of agitation provided by the stirrer. Table 7.1 lists some operating values for experiments.

The following steps are used to obtain the experimental data using the virtual experiment.

1. From the introductory screen **Laboratory Overview,** select appropriate buttons to view **Overview, Industrial Systems, Theory, and Procedures** that describe various aspects of the experiment. Then, select the button for **Virtual Experiment** to conduct the experiment.
2. In the **Control panel** (Fig. 7.1), enter the selected values for initial temperature (°C), kettle surface temperature (°C), radius of kettle (m), heating time (min), and the convective heat transfer coefficient (W/m²-K).
3. After entering all the variables, select **Start Experiment** to begin the experiment.

Table 7.1. Suggested process variables.

Trial	Heat Transfer Coefficient (W/m²-K)	Kettle Surface Temperature (°C)
1	1000	90
2	2500	100
3	5000	110
4	2500	90
5	5000	100
6	1000	110
7	5000	90
8	1000	100
9	2500	110

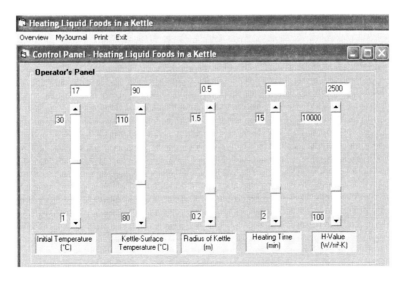

Figure 7.1. Control panel to determine the temperature change of tomato juice heated in a hemispherical, steam-jacketed kettle.

Heating Liquid Foods in a Kettle

Overview   MyJournal   Print   Exit

Data

| A16 | 9.639 | | | | |

	A	B	C	D	E	F
1	Exp.No 1		Exp.No 2		Exp.No 3	
2	Ti(°C)=		17 Ti(°C)=		17 Ti(°C)=	17
3	Tkettle(°C)=		90 Tkettle(°C)=		90 Tkettle(°C)=	90
4	Radius(m)=		0.5 Radius(m)=		0.5 Radius(m)=	0.5
5	Time(Min)=		300 Time(Min)=		300 Time(Min)=	300
6	H(W/m²·K)=		2500 H(W/m²·K)=		4000 H(W/m²·K)=	6000
7	Time(s)	T(°C)	Time(s)	T(°C)	Time(s)	T(°C)
8	0	17	0	17	0	17
9	1.205	17.34	1.205	17.543	1.205	17.813
10	2.41	17.678	2.41	18.082	2.41	18.618
11	3.614	18.015	3.614	18.618	3.614	19.413
12	4.819	18.351	4.819	19.149	4.819	20.2
13	6.024	18.684	6.024	19.676	6.024	20.977
14	7.229	19.016	7.229	20.2	7.229	21.746
15	8.434	19.347	8.434	20.719	8.434	22.507

Figure 7.2. The experimental data collected for different trials.

7.3 Results:
After the **Start Experiment** button is selected, a plot of temperature with respect to time is shown. Repeat the trials with other conditions given in Table 7.1. When the **View Data in Spreadsheet** button is clicked at the end of all trials, all the experimental data for different trials will be shown on the spreadsheet. Save the data in Excel (*.xls) format (Fig. 7.2).

7.4 Analysis of Results:
In this experiment, you obtained temperature-time data for heating tomato juice in a hemispherical, steam-jacketed kettle for different processing conditions. Using these data:

1. Create plots of temperature versus time for different kettle surface temperatures and constant heat transfer coefficients.

2. Determine the heating rate using the following steps:

   - In a new column, using the temperature-time data, determine $\ln\left[\dfrac{T-T_m}{T_i-T_m}\right]$ versus time.

   - Apply the **Trendline** feature of Excel to determine the slope of the linear portion of the above curve.

- Determine the heating rate (1/s) from the slope using the following expression:

$$Heating\ rate = -(slope) \qquad\qquad 7.3$$

7.5 Discussion:

1. Does juice temperature increase with increasing convective heat transfer coefficient at any constant kettle surface temperature? Why?
2. What is the role of kettle surface temperature and convective heat transfer coefficient on the heating process?
3. Why is the assumption of lumped capacity valid for heating juice in a kettle?

7.6 Online Links:

Groen Steam-Jacketed Kettles:
    http://www.groen.com
Five Star Marketing Inc.:
    http://www.fivestarmktg.com/floorkettles.asp
Market Forge Industries Inc.:
    http://www.mfii.com/products/food/steam_jacketed_kett le/specsheets/
Fesmag Food Service Equipment:
    http://www.fesmag.com

7.7 Bibliography:

Lewis, M.J. 1987. Physical Properties of Foods and Food Processing Systems. Ellis Horwood Ltd., Chichester, England.

Singh, R.P. and Heldman, D.R. 2001. Introduction to Food Engineering. Academic Press. London.

Chapter 8 Canning Foods – Determining Heat Rate Parameters in Conduction-Heating of Foods in a Can

Thermal processing is one of the most common methods used in preserving foods for long duration. Heating foods to temperatures that destroy pathogenic microorganisms is effective in making foods safe for human consumption. Heating is also beneficial in destroying spoilage organisms and enzymes that cause foods to spoil. Heating alters some sensory and textural attributes of foods, which may be either desirable or undesirable. Therefore, thermal processes (temperature and time of heating) are carefully chosen to obtain maximum benefit of extended shelf life with minimal adverse effects on the quality attributes of foods.

In commercial thermal processing, canning and aseptic processing are two different methods commonly used. In canning, first the food is hermetically packed in a metal can or glass bottle, and then it is sterilized by applying heat to destroy any microorganisms and/or their spores and to inactivate enzymes (Fellows, 2000; Heldman and Hartel, 1997). The equipment used for the sterilization of canned (or bottled) foods is called a retort.

To evaluate the effectiveness of thermal processing, it is important to know the thermal history of the product, the thermal resistance of the microorganisms of concern, and the quality attributes of the food.

In this laboratory exercise, we will first view the different types of industrial systems. Then we will conduct a virtual experiment to sterilize a canned food. We will be able to obtain the thermal history of the product at different locations inside a can. The data analysis will include calculating different heating rate parameters to determine the effectiveness of the process. Specifically, we will determine:

- the slowest heating point by plotting the time-temperature data obtained from different locations in the can, and
- heating (f_h) and cooling (f_c) rates at desired locations within the can.

8.1 Objectives:
1. To determine the temperature history of a canned food subjected to thermal processing in a retort.
2. To determine the heating (f_h) and cooling (f_c) rates at different locations within the can.

3. To evaluate the change of heating (f_h) and cooling (f_c) rates with respect to different retort temperatures, locations within the can, and can sizes.

8.2 Materials and Methods:

In a laboratory experiment to obtain the thermal history inside a can, thermocouples are inserted at three different locations within the can. Special fittings are used to make sure that the insertion openings through the lid are properly sealed with rubber gaskets. Select **Procedures** to view how to insert thermocouple connectors into the lid of a can. After the diameter and height of the can are measured with a caliper, the can is fitted with thermocouple sensors and placed in a retort.

For the purpose of this experiment, three different can sizes have been selected, 211×304, 307×409, and 401×411. Using a given can dimension, the heating (f_h) and cooling (f_c) rates at three different locations within the can will be calculated, and the effect of can size and location of temperature measurement on f_h and f_c will be determined. For this experiment, the initial temperature of the canned food will be kept constant at 25°C.

Typical control equipment available on commercial retorts allows setting desired steam and cooling water temperatures and total heating and cooling times. In this experiment, three different steam temperatures (110, 121 and 130°C) will be used, and the cooling water temperatures will be constant at 20°C. A heating time of 40 min will be used. A suggested list for can dimensions and retort temperatures is shown in Table 8.1.

Table 8.1. Suggested product and process variables.

Trial	Can Size	Retort Temperature (°C)
1	211×304	130
2	307×409	110
3	401×411	121
4	211×304	121
5	307×409	130
6	401×411	110

The following steps are used to obtain the experimental data using the virtual experiment.

- From the introductory screen **Laboratory Overview,** select appropriate buttons to view **Overview, Industrial Systems, Theory, and Procedures** that describe various aspects of the experiment. Then select the button for **Virtual Experiment** to conduct the experiment.
- Select the size of can. The dimensions will be given in appropriate text boxes. You may change the dimensions to any value of interest by entering new values in these text boxes.
- From the **Control Panel**, select the process variables such as *Initial Temperature, Retort Temperature, Cooling Temperature*, and *Heating Time* using the vertical scroll bars (or directly input the required values in the appropriate text boxes). Click on the **Start Experiment** to begin the experiment and collect the data (Fig. 8.1). Note that in the virtual experiment the cooling time is equal to the heating time. After completing the experiments using suggested conditions (Table 8.1), you may try additional experiments using any other desired operating conditions.

8.3 Results:

After the **Start Experiment** button is selected, the temperature change at three different locations of the can (T-1: center point, T-2: midway between the center and the corner, and T-3: corner) is shown in a plot, and a visual representation of the cross-section of the can is shown lengthwise. A color change takes place as the heat transfers into the product. Initially, the temperature distribution is uniform (as shown with blue color). As heating begins and continues from the surface to the inside, the heating front (as shown with red color) moves toward the center (Fig. 8.2), colors reverse after the cooling starts. The process time is shown in the text boxes under the plot (Fig. 8.2).

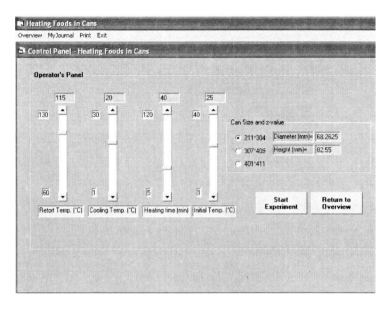

Figure 8.1. Control panel to select the process and product variables for the experiment.

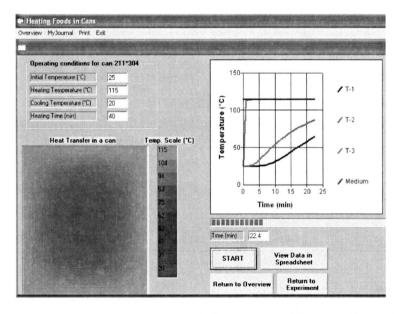

Figure 8.2. Center temperature change and color-coordinated picture of central cross-section with temperature during heating.

Once the experiment for the selected conditions in the previous screen is completed, you will need to save the data for further analysis. Save the data by selecting the **View Data in Spreadsheet** button. Save by selecting the **Save as...** button in an appropriate directory in Excel (*.xls) format.

8.4 Analysis of Results:

In this experiment, you obtained the temperature-time data for the various operating conditions. The **Data Analysis** button may also be used to view the steps for data analysis.

Using temperature-time data:

1. Create plots of temperature vs. time for different locations inside the can.
2. Compare the plots to determine the slowest heating location in the can.

Determine the heating and cooling rates at these locations following the procedure given below:

The governing equations for a typical heating or cooling curve may be written as follows:

$$\textit{Heating phase}: \log\left(T_m - T\right) = -\frac{1}{f_h}t + \log\left(T_m - T_{pih}\right) \qquad 8.1$$

$$\textit{Cooling phase}: \log\left(T - T_w\right) = -\frac{1}{f_c}t + \log\left(T_{pic} - T_w\right) \qquad 8.2$$

where T_m is the retort temperature for heating (°C), T is product temperature at a certain location (°C), T_w is cooling water temperature (°C), T_{pih} is pseudo initial heating temperature (°C), T_{pic} is pseudo initial cooling temperature (°C), f_h is heating rate factor (s), f_c is cooling rate factor (s), j_h is heating lag, and j_c is cooling lag.

Using the experimentally obtained temperature-time data and the characteristics of these equations, the heating and cooling rates may be calculated as follows:

1. Copy the temperature time data in an Excel spreadsheet,
2. In a new column determine the $\log(T_m-T)$, (Fig. 8.3),
3. Create a scatter plot of $\log(T_m-T)$ versus time (Fig. 8.3),
4. Find the linear portion of the plot visually (in Fig. 8.3, the linear part of the plot starts from t ≈ 20min). Delete the nonlinear portion of the data from the data columns, and redraw the plot.
5. Apply the **Trendline** feature of Excel for the log (T_m-T) versus time data in the visually determined linear portion to calculate the f_h and T_{pih} values.
6. When the **Trendline** procedure is applied to the linear portion of the plot, the intercept equals $\log(T_m-T_{pih})$ and the slope is $-1/f_h$.

7. Calculate the f_h and T_{pih} values (Fig 8.4).

$$T_{pih} = T_m - 10^{(int\,ercept)} \qquad\qquad 8.3$$

$$f_h = -\frac{1}{slope} \qquad\qquad 8.4$$

8. Calculate the j_h value using Eq. 8.3 (Fig 8.4)

$$j_h = \frac{T_m - T_{pih}}{T_m - T_i} \qquad\qquad 8.5$$

where T_i is the initial product temperature, (°C).

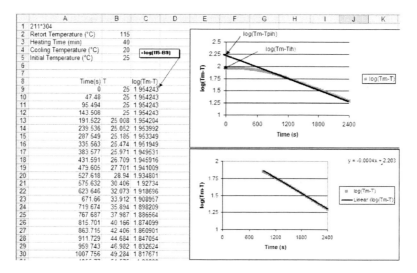

Figure 8.3. Time-temperature data collected from a heating trial

	E	F	G	H	I	J	K
32	Slope of the fitted line (1/s)=			-0.0004	=-1/H32		
33	fh (s) =			2500			
34	log(Tm-Tpih)=			2.203	=10^H34		
35	(Tm-Tpih)=			159.6	=115-H35		
36	Tpih=			-44.6	=115-H36		
37	jh=			1.8	=(115-H36)/(115-25)		

Fig 8.4 Calculation of heating rate parameters

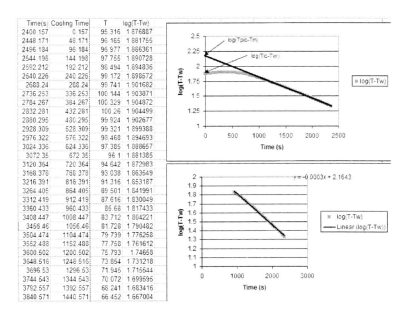

Time(s)	Cooling Time	T	log(T-Tw)
2400.157	0.157	95.316	1.876887
2448.171	48.171	96.165	1.881755
2496.184	96.184	96.977	1.886361
2544.198	144.198	97.755	1.890728
2592.212	192.212	98.494	1.894836
2640.226	240.226	99.172	1.898572
2688.24	288.24	99.741	1.901682
2736.253	336.253	100.144	1.903871
2784.267	384.267	100.329	1.904872
2832.281	432.281	100.26	1.904499
2880.295	480.295	99.924	1.902677
2928.309	528.309	99.321	1.899388
2976.322	576.322	98.468	1.894693
3024.336	624.336	97.385	1.888657
3072.35	672.35	96.1	1.881385
3120.364	720.364	94.642	1.872983
3168.378	768.378	93.038	1.863649
3216.391	816.391	91.316	1.853187
3264.405	864.405	89.501	1.841991
3312.419	912.419	87.616	1.830049
3360.433	960.433	85.68	1.817433
3408.447	1008.447	83.712	1.804221
3456.46	1056.46	81.728	1.790482
3504.474	1104.474	79.739	1.776258
3552.488	1152.488	77.758	1.761612
3600.502	1200.502	75.793	1.74658
3648.516	1248.516	73.854	1.731218
3696.53	1296.53	71.945	1.715544
3744.543	1344.543	70.072	1.699595
3792.557	1392.557	68.241	1.683416
3840.571	1440.571	66.452	1.667004

Fig 8.5 Time temperature data collected for the cooling trial

	K	L	M	N	O	P	Q
45	Slope of the fitted line (1/s)=			-0.0003	=-1/N45		
46	fc (s)=			3333.3			
47	log(Tpic-Tw)=			2.1643	=10^N47		
48	(Tpic-Tw)=			145.98			
49	Tpic=			165.98	=N48+20		
50	jc=			1.938	=(N49-20)/(95.316-20)		
51							

Fig 8.6 Calculation of the cooling parameters

The cooling rate parameters are obtained from the cooling data. An extra precaution must be taken while applying the above procedure to the cooling data. Since cooling data are recorded together with the heating data, the cooling times must be re-calculated, subtracting the heating time, at the end of heating cycle, from the elapsed time as shown in Fig. 8.5. The cooling rate parameter, f_c, is

$$f_c = -\frac{1}{slope} \qquad (8.6)$$

and, the pseudo initial cooling temperature, T_{pic}, is obtained as

$$T_{pic} = T_w + 10^{intercept} \qquad (8.7)$$

and parameter j_c is obtained as follows:

$$j_c = \frac{T_{pic} - T_w}{T_{ic} - T_w} \qquad (8.8)$$

where T_{ic} is the temperature of product at the beginning of cooling cycle.

The cooling rate parameters are calculated as shown in Fig. 8.6. Repeat these calculations for each temperature-time data obtained from the experiment for different locations for both heating and cooling phases.

8.5 Discussion:
1. How are the f_h and f_c values affected by location and time? Discuss the reasons.
2. Does the j-value change by location or time? Discuss the reasons.
3. Where would you expect the slowest heating point in a can of conduction heated food to be located? Discuss the reasons.
4. Can f- and j-values be used to determine the time required for the slowest heating point in the can to reach a selected temperature? Discuss.

8.6 Online Links:
Steritech Steam Retorts:
> http://www.steritech-fr.com

Stork Food and Dairy Systems Inc.:
> http://www.stork-usa.com/Retorts.htm

Lagarde Autoclaves:
> http://www.lagarde-autoclaves.com/html/gb/index-agro-gb.htm

Stock America:
> http://www.stockamerica.com/retort.html

Dixie Canner Corporation-Can Packaging and Processing Equipment:
> http://www.dixiecanner.com/

Alard Equipment Corporation-Supplying food processing machinery
> http://www.alard-equipment.com/genlistings.htm#R

Allpax-Process Equipment for Food and Pharmaceuticals Manufacturers:
> http://www.allpax.com
> http://www.allpax.com/steam.htm
> http://www.allpax.com/spray.htm
> http://www.allpax.com/immersion.htm
> http://www.allpax.com/steamair.htm
> http://www.allpax.com/verticals.htm

http://www.allpax.com/horizontal.htm
Ecklund:
http://www.tcal.com
FMC FoodTech:
http://www.fmcfoodtech.com
Temco Incorporation:
http://www.temco.com/retort.htm
Selas Corporation of America-The Heat Technology Company:
http://www.selas.com/retort.htm

8.7 Bibliography:

Fellows, P.J. 2000. Food Processing Technology: Principles and Practice. CRC Press, Boca Raton, FL.

Heldman, D.R. and Hartel, R.W. 1997. Principles of Food Processing. Chapman & Hall. New York, N.Y.

Karel, M., and Lund, D.B. (2003). "Physical Principles of Food Preservation," 2nd ed., Marcel Dekker, Inc., New York.

Singh R.P. and Heldman, D.R. (2001). "Introduction to Food Engineering,: 3rd. ed., Academic Press, London.

Toledo, R.T. (1994). "Fundamentals of Food Process Engineering," 2nd ed., Chapman and Hall, New York.

Chapter 9 Lethality of a Thermal Process – Determining Lethality during Heating of a Canned Food

The primary objective of thermal processing of foods is to ensure the safety of the product. The lethality of microorganisms obtained during canning depends on the heating conditions, the heat resistance of the microorganisms and the heating characteristics of a food product. The decimal reduction time, or D-value, is the time required to reduce the population of a microorganism by one log-cycle (or by 90%):

$$D = \frac{t}{\log(N_0) - \log(N)} \qquad 9.1$$

where t is the time, N_0 and N are initial and final number of microorganisms, respectively (Singh and Heldman, 2001). The effect of temperature on the D-value is called the thermal resistance factor known as the z-value. The z-value is unique for each microorganism depending on the surrounding medium. It may be defined as the temperature change required to reduce the D-value by 90%.

The total time required to accomplish a certain reduction in the number of microorganisms is called the thermal death time (F value):

$$F = D \log\left(\frac{N_0}{N}\right) \qquad 9.2$$

The F-value of an applied thermal process is often required, since it indicates the number of log cycles of microbial reduction (Singh and Heldman, 2001). The F-value is generally calculated based on knowing the temperature history of the product at its coldest point, as given in Eq. 9.3:

$$F = \int_0^t 10^{\frac{T_c(t) - T_r}{z}} \, dt \qquad 9.3$$

where $T_c(t)$ is the temperature at the slowest heating point as a function of time, T_r is the reference temperature (121.1°C), z is the z-value of the microorganism, and t is the processing time.

In the experiment described in Chapter 8, heat penetration characteristics of the conduction heating foods were studied. In this laboratory exercise, we will again conduct experiments using a retort to obtain the time-temperature history of a canned food product. The data analysis will include determining the F-value for different microorganisms, using the temperature profiles at different locations in a can and processed at different retort temperatures.

9.1 Objectives:

1. To experimentally determine the temperature history of a canned food subjected to different retort temperatures.
2. To determine the F-value at the slowest heating point and other locations in a can for different microorganisms.
3. To examine the effect of retort temperature and different locations in a can on the resulting F-values.

9.2 Materials and Methods:

In a laboratory experiment, the can size is measured using calipers. Using special attachments, thermocouples are inserted in the can so that the measuring junction is located at the desired location. Inoculated conduction-heating food is filled in the cans and the cans are processed in a batch retort. After a given thermal process, the lethality is determined.

For the purpose of the virtual experiment, one can size (307x409) has been selected. Using the given dimensions of the can, the center temperature profile is obtained for different retort temperatures. The F-value is calculated using different z-values (5, 10, and 20°C). The initial temperature of the canned product is kept constant at 25°C. You may try different sets of trials at other initial temperatures to determine if the initial temperature has any effect on the resulting F-value.

In this experiment, the retort temperatures of 121.1 and 130°C are used. The initial temperature of the product and the cooling water temperature is set at 20°C, and 60 min of heating time is used for different trials. Table 9.1 lists the trials to be used in this experiment.

This experiment is similar to the heat penetration experiment given in the previous chapter. However, you need to select a z-value in the control panel (Fig. 9.1). The following steps are used to run the experiment.

Table 9.1. Suggested product and process variables.

Trial	z-value	Retort Temperature (°C)
1	10	121.1
2	20	130
3	5	121.1
4	10	130
5	5	130
6	20	121.1

1. From the introductory screen **Laboratory Overview**, select buttons to view **Overview, Industrial Systems. Theory** and **Procedures** for visual and text descriptions of industrial-type retorts and procedures to be used during the experiment. Then, select the button for **Virtual Experiment**.

2. From the **Control Panel**, enter desired values for *Retort Temperature, Cooling Temperature,* and *Heating Time* using the vertical scroll bars (or directly enter the required values in the appropriate text boxes). Click on **Start Experiment** to start the experiment and collect the data (Fig. 9.1).

9.3 Results

Once the experiment for the selected conditions is completed, you may need to save the data for further analysis. To save the data, click on the **View Data in Spreadsheet** button. Then you may save it by selecting the **Save as...** button in an appropriate directory in any format you choose, preferably in Excel (*.xls). After the data are saved, you may return to the **Virtual Experiment** screen for a new set of conditions for product and process variables.

9.4 Analysis of Results:

In this experiment, you obtained the time-temperature data for a given can size of 307x409 for different retort temperatures at different locations in the can (T-1: center point, T-2: midway

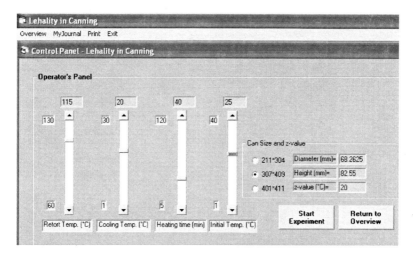

Figure 9.1. Control panel to select the process and product variables for the experiment.

between the center and the corner, and T-3: corner). The "**Data Analysis**" option may also be used to view the steps for data analysis.

Using time-temperature data create plots of temperature profiles for the different locations, and compare the plots to determine the coldest point. The saved data in Excel format for any trial will appear as seen in Fig. 9.2. The reported F-value in this figure (Cell B6) is for a trial when z = 20°C.

After plotting the temperature profiles for the given locations and determining the one for the coldest point, you may calculate the F-values to compare with those predicted in the experiment.

The following procedure applies for the slowest heating point for the data given in Fig. 9.2 (Column B):

1) Copy columns A and B of time-temperature data into a new worksheet (Fig. 9.3).
2) To determine the F-value using Eq. 9.3, first we will calculate
    $$L = 10^{\frac{T(t)-T_r}{z}}$$
    for z =20 °C (Fig. 9.3 column C).
3) Create a plot of L values (column C) and time as shown in Fig. 9.4. Next, we need to determine the area under the L-curve by integration.
4) The integration step is accomplished using the trapezoidal rule, where the area under the L-value curve will be assumed to consist of numerous trapezoids:
    *   The change in L-value between two times will be the longer and shorter dimensions of the trapezoid, and the difference between these two time locations will be the height of the trapezoid (Fig. 9.4).

- The area of a trapezoid is given by:

$$A = \frac{a+c}{2}h \qquad\qquad 9.4$$

	A	B
1	307*409	
2	Retort Temperature (°C)	115
3	Heating Time (min)	60
4	Cooling Temperature (°C)	20
5	Initial Temperature (°C)	25
6	Fo center (min)	6.15
7	See text for T-1, T-2, T3 locations.	
8	Time(s)	T-1
9	0	25
10	35.744	25
11	72.021	25
12	108.298	25
13	144.575	25.001
14	180.852	25.005
15	217.129	25.024
16	253.406	25.078
17	289.683	25.194
18	325.96	25.401
19	362.237	25.721
20	398.515	26.174
21	434.792	26.768
22	471.069	27.507
23	507.346	28.388

Figure 9.2. Time-temperature data saved in Excel format.

	A	B	C	D	E	F	G
1		=10^((B2-121.1)/20)		=((C4+C5)*(A5-A4))/2			
2						=SUM(D4:D104)	
3	Time(s)	T-1		Integration Column			
4	0	25	1.56675E-05	0.00056002	F=	369.4 seconds	
5	35.744	25	1.56675E-05	0.00056837	F=	6.1567 min	
6	72.021	25	1.56675E-05	0.00056837			
7	108.298	25	1.56675E-05	0.000568403		=F4/60	
8	144.575	25.001	1.56693E-05	0.000568567			
9	180.852	25.005	1.56765E-05	0.00056932			
10	217.129	25.024	1.57109E-05	0.00057172			
11	253.406	25.078	1.58088E-05	0.000577352			
12	289.683	25.194	1.60214E-05	0.000588216			
13	325.96	25.401	1.64078E-05	0.000606394			

Figure 9.3. Integrating time-temperature data to determine the F-value.

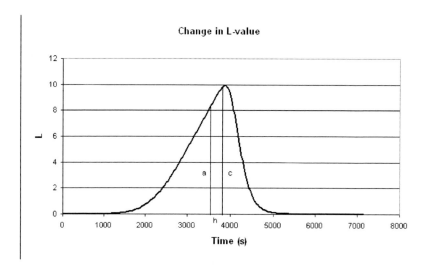

Figure 9.4. Application of the trapezoid rule to determine the F-value.

The area under the L-value curve will be the summation of the areas of all trapezoids (Fig. 9.4), which will result in the F-value in seconds as shown in cell F4 (Fig. 9.3): the summation of the values given in column D. We may then compare our results with those obtained in the virtual experiment (cell B6 in Fig. 9.2).

9.5 Discussion:
   1. How does the F-value change by retort temperature? Does it change by location in the can?
   2. What is the effect of z-value on F-value? Does F-value increase with an increase in z-value?
   3. Determine the accomplished log cycle reductions in your trials for a D-value of 2 min?

9.6 Online Links:
Steritech Steam Retorts:
        http://www.steritech-fr.com
Stork Food and Dairy Systems Inc.:
        http://www.stork-usa.com/Retorts.htm
Lagarde Autoclaves:
        http://www.lagarde-autoclaves.com/html/gb/index-agro-gb.htm
Stock America:
        http://www.stockamerica.com/retort.html
Dixie Canner Corporation-Can Packaging and Processing Equipment:
        http://www.dixiecanner.com/

Alard   Equipment   Corporation-Supplying   food   processing machinery

http://www.alard-equipment.com/genlistings.htm#R

Allpax-Process   Equipment   for   Food   and   Pharmaceutical Manufacturers:

http://www.allpax.com

9.7 Bibliography:

Karel, M. and D.B. Lund. 2003. Physical Principles of Food Preservation. 2nd Edition. Marcel Dekker. New York.

Singh, R.P. and Heldman, D.R. 2001. Introduction to Food Engineering. Academic Press. London

Chapter 10 Hydrocooling – Cooling Melons and Cherries in a Hydrocooler

Fruits and vegetables are commonly harvested in the summer months, when the day temperatures in many growing regions reach 35°C and above. The temperature of the harvested product is usually in equilibrium with the ambient air. Several metabolic reactions in the harvested product occur quickly, causing irreversible changes that impair the quality of the product. Prompt lowering of the temperature of the harvested product can reduce the rates of these undesirable reactions. Depending upon the product, several methods are used for cooling harvested fruits and vegetables. The most common procedures include immersing the product in chilled water or spraying chilled water on the product. The equipment used for this purpose is called a hydrocooler.

In this laboratory exercise, we will conduct an experiment with a hydrocooler that involves immersing foods in chilled water. We will use melons and cherries of different sizes to obtain temperature histories at the product center and mass average temperature vs. time. Data analysis will include calculating a commonly used industrial term, "7/8th cooling time." In commercial practice, fruits and vegetables often are cooled to a temperature representing 7/8th cooling time.

10.1 Objectives:

1. To determine the temperature history at the geometric center of melons and cherries when immersed in chilled water during hydrocooling.
2. To determine the mass average temperature vs. time profiles for melons and cherries of different sizes during hydrocooling.
3. To examine the effect of melon and cherry size on the cooling rate and 7/8th cooling time.
4. To determine the effect of water temperature and velocity on the cooling times for cherries or melons of a given size.

10.2 Materials and Methods:

We have selected melons and cherries for this experiment. Both fruits represent the shape of a sphere. We will use a hydrocooler to immerse products in clean, potable water chilled to low temperature. While heat will be transferred from the product being cooled to the chilled water, the water will be cooled using a heat exchanger located outside the hydrocooler. Chilled water will

Table 10.1   Suggested conditions for virtual experiments

Trial	Melon Diameter (mm)	Cherry Diameter (mm)	Water Temperature (°C)
1	-	20	2
2	-	25	2
3	-	30	2
4	-	25	1
5	-	25	2
6	-	25	3
7	100	-	2
8	130	-	2
9	160	-	2

be circulated through the hydrocooler tank to obtain a constant temperature in the tank.

The following steps will be used to conduct the virtual experiments.

1. From the introductory screen **Laboratory Overview**, select buttons to view **Overview, Industrial Systems, Theory, and Procedures** to view visual and text description of industrial hydrocoolers. Then, select the button for **Virtual Experiment**.
2. Select the product of interest: melon or cherries (Fig 10.1).
3. After selecting the process variables (**Initial Temperature, Final Center Temperature, Water Temperature,** and **Water Velocity**) using the vertical scroll bars (or directly entering the text boxes), click on the **Start Experiment** button to start the experiment and collect the data.

10.3 Results:

After the **Start Experiment** button is selected, you will see a cross-section of the product. As heat transfers out of the product, the color of the cross-section will change with temperature (Figure 10.2). Initially, the temperature throughout the cross-section of the product will be uniform, as shown in red. As cooling begins, the circumferential region will begin to cool as denoted by shades of blue. As cooling progresses, the cooling front moves towards the center. On the left side of the screen, a graph (Figure 10.2) will be shown where the temperature at the product center, product mass average, and water are plotted.

Once the experiment for the selected conditions is completed, you will need to save the data for further analysis. Click on the **View Data in Spreadsheet** button. You will see the spreadsheet screen (Figure 10.3). The spreadsheet contains the temperature time data for the center, midpoint, and mass average. To save these data, in Excel format follow the procedures given in Chapter 1. After saving the data, you may return to the "**Virtual Experiment**" screen, and a new set of conditions for product and process variables may be selected for another experiment.

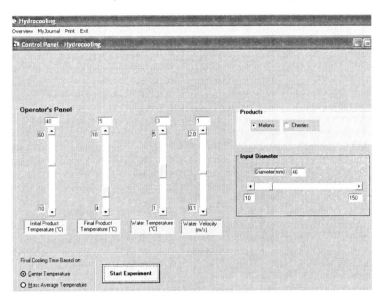

Figure 10.1. Virtual experiment, product and process variables control screen.

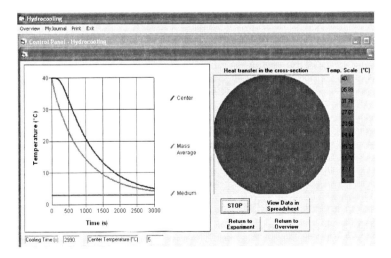

Figure 10.2 A representation of center temperature change and color-coordinated picture of the product cross-section with temperature during hydrocooling.

Data0

A18 | 240

	A	B	C
1	Initial Temperature (°C)	40	
2	Water Temperature(°C)	3	
3	Water Velocity (m/s)	1	
4	Product:	Melon	
5	Time (s)	Center Temperature (°C)	Mass Average Temperature (°C)
6	0	40	40
7	20	40	38.878
8	40	40	37.872
9	60	40	36.931
10	80	40	36.041
11	100	39.997	35.191
12	120	39.985	34.378
13	140	39.953	33.596
14	160	39.889	32.843
15	180	39.784	32.116
16	200	39.632	31.414
17	220	39.429	30.733
18	240	39.176	30.074

Fig 10.3 Data collected in the hydrocooling experiment.

After the data are saved, you may return to the **Virtual Experiment** screen (by clicking on it), and a new set of conditions for product and process variables may be selected for the new experiment.

10.4 Analysis of Results:

In this experiment, you obtained temperature-time data for products of different size, cooled with different water temperatures (Table 10.1). For each set of experimental variables, the results should have been saved on spreadsheet files.

Use the following steps to analyze the results (click on the **Data Analysis** button to view steps for data analysis).

Using temperature-time data (Fig. 10.3), create plots of center and mass average temperature vs. time for each set of experimental variables. Observe the effect of size, water temperature, and product on temperature decrease. In addition, calculate the 7/8th cooling time as follows.

7/8th cooling time

A common parameter used in the industry to describe the cooling rate of fresh fruits and vegetables is the 7/8th cooling time. The procedure to calculate 7/8th cooling is as follows:

1. From the plot of mass average temperature vs. time, determine the time when temperature reaches half-cooling

or $T_{1/2} = (T_o - T_m)/2$,

where T_o is the initial product temperature, and T_m is cooling medium temperature.

2. The 7/8th time is three times the half-time for cooling. (Note: 7/8th cooling time is calculated only from mass average temperature data.)

Example

1. For the temperature-time data obtained on the spreadsheet (Figure 10.3) for a 46 mm melon, the initial temperature is 40°C, and the cooling medium temperature is 3°C. The time required to reduce the temperature by half the difference between initial product temperature and cooling medium is (40-3)/2 = 18.5°C.
2. Identify 18.5°C on the mass average temperature column, and determine the corresponding time. In this example, it is 720 s or 12 min.
3. Multiply the time obtained in Step 2 by 3 to obtain 7/8th time. For this example, the answer is 36 min.

10.5 Discussion:

1. Discuss how the product size influences the 7/8th cooling time.
2. Why are the profiles of center temperature and mass average temperature different for any given product size?
3. Discuss the role of cooling water temperature on the 7/8th cooling time.

10.6 Online Links:

Peal-engineering:
>   http://www.peal-engineering.co.uk/

Norlock Refrigeration
>   http://www.norlockrefrigeration.com/index.html

Ekko
>   http://www.ekkoas.dk/

Horticold:
>   http://www.horticold.com/index.html

10.7 Bibliography:

Thompson, J.F., F.G. Mitchell, T.R. Rumsey, R. Kasmire, and C.H. Crisosto. 1998. Commercial Cooling of Fruits, Vegetables, and Flowers. Publication 21567, University of California, Division of Agricultural and Natural Resources, Oakland, California

Kasmire, R.F., and R.A. Parsons. 1971. Precooling cantaloupes, a shipper's guide. Agric. Ext. Service, University of California,

Berkeley.

ASHRAE 1990. American Society of Heating, Refrigeration and Air Conditioning Engineers (ASHRAE). Guide and Databook. Atlanta: ASHRAE.

Singh, R.P. and Heldman, D.R. 2001. Introduction to Food Engineering. Academic Press. London.

Chapter 11 Kinetics of Nutrient Degradation – Determining Kinetics of Ascorbic Acid Loss during Heating of Orange Juice

Thermal processing of foods is based on the destruction of microorganisms to ensure safety. However, thermal processing also results in some undesirable changes, such as sensory (e.g., discoloration, flavor and textural changes) and physical and chemical changes (e.g., over-cooking, liquefaction, and nutrient loss). Thus, thermal processing of foods has two opposing effects that are both time- and temperature-dependent. Most nutrients are much more resistant to heat compared with the microorganisms, so it is possible to minimize nutrient loss while ensuring the safety of the product. However, the kinetic data (E_a and k_T) to describe the temperature-time effects on a nutrient should be known to achieve this.

Kinetic change of a quality factor may be described by:

$$\frac{dQ}{dt} = -k_T Q^n \qquad 11.1$$

where k_T is reaction rate constant, Q is the nutrient concentration, t is time, and n is reaction order.

Another common parameter used in thermal process calculations is the D-value. For a microorganism, the D-value is the time at a certain temperature that results in 90% reduction of its initial population. The relationship between the k-value and D-value is:

$$k_T = \frac{2.303}{D_T} \qquad 11.2$$

The dependence of k on temperature may be given by Arrhenius equation:

$$k_T = Ae^{\left(-\frac{E_a}{RT}\right)} \qquad 11.3$$

where E_a is activation energy (cal/mol), R is ideal gas constant (1.987 cal/molK), T is temperature (K), and A is the pre-exponential constant.

Ascorbic acid (vitamin C), is an important nutrient in the human diet. Fruits and vegetables are rich in their ascorbic acid content. Since it is a thermolabile substance, thermal processing of fruits and vegetables may significantly decrease the amount of ascorbic acid.

In this laboratory, we will conduct an experiment to determine the kinetics of ascorbic acid degradation in orange juice during thermal processing. We will heat orange juice to different temperatures and determine the change in ascorbic acid at different times. Data analysis will include calculating the kinetic parameters, such as k_T- and E_a-values.

11.1 Objectives:
1. To determine the change in ascorbic acid in samples of orange juice heated to different temperatures.
2. To determine the rate constants at different temperatures using data on change in the ascorbic acid concentration.
3. To determine the E_a (activation energy) values at different processing temperatures.

11.2 Materials and Methods:
We have selected orange juice for this experiment. A retort is used to process the samples at different temperatures (90, 100 and 110°C), and the ascorbic acid amount in the samples is experimentally determined at selected time intervals. These data are reported as Q/Q_0 value of the ascorbic acid, and Eqs. 11.1 to 11.3 are used to determine the k-, D-, and E_a values.

The following steps are used to obtain the experimental data using the virtual experiment.

From the introductory screen **Laboratory Overview,** select the appropriate buttons to view **Overview, Industrial Systems, Theory, and Procedures** that describe various aspects of the experiment. Then, select the button for **Virtual Experiment** to conduct the experiment.

After selecting the process variable (temperature) using the horizontal scroll bar, click on the **Start Experiment** button to begin the experiment and collect the data (Fig. 11.1). Use suggested process variables as shown in Table 11.1.

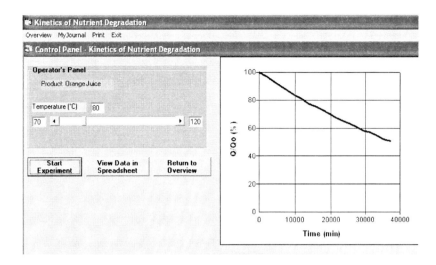

Figure 11.1. Control screen for the virtual experiment to select product and process variables.

Table 11.1 Suggested process variables

Product	Temperature (°C)
Orange Juice	90
	100
	110

11.3 Results:

After the **Start Experiment** button is selected, a plot of (Q/Q_0) of ascorbic acid concentrations vs. time will be shown (Fig. 11.1). Once the experiment for the different processing temperatures is completed, click on the **View Data in Spreadsheet** button. Save the data by selecting the **Save as...** button in an appropriate directory in Excel format (*.xls). The experimental data consists of the number of experiments, processing temperature, and the Q/Q_0 vs. time data.

11.4 Analysis of Results:

	A	B	C	D	E	F
1	Exp.:1		Exp.:2		Exp.:3	
2	T(°C)=	80	T(°C)=	90	T(°C)=	100
3	t(min)	Q/Qo	t(min)	Q/Qo	t(min)	Q/Qo
4	0	1	0	1	0	1
5	1965.883	0.971	591.093	0.968	189.537	0.967
6	3931.766	0.935	1182.187	0.935	379.073	0.935
7	5897.649	0.902	1773.28	0.904	568.61	0.897
8	7863.532	0.867	2364.373	0.865	758.147	0.874
9	9829.415	0.836	2955.466	0.839	947.684	0.84
10	11795.3	0.811	3546.56	0.808	1137.22	0.804

Figure 11.2. Experimental data obtained for different processing temperatures for orange juice.

The reaction order (n) for ascorbic acid degradation has been reported as 1 in the literature. To analyze results of this experiment, we will also use n=1. In this experiment, you obtained the Q/Q_0 vs. time data of ascorbic acid for orange juice subjected to different thermal processing temperatures. For each set of experimental variables, the results should have been saved in the same spreadsheet file. "**Data Analysis**" option may also be used to view the steps for data analysis.

The following steps are used to determine rate constants and activation energy.
1) Convert Q/Q_0 into $\ln(Q/Q_0)$ as shown in Fig. 11.3. Create plots of $\ln(Q/Q_0)$ versus time for different temperatures (Fig. 11.4).
2) Determine the slope (Fig. 11.4) using the **Trendline** procedure in Excel.
3) Eq. 11.1 may be re-written as:

$$\ln\left(\frac{Q}{Q_0}\right) = -k_T t \qquad 11.4$$

From the plot of ln (Q/Q_0) vs. t determine the slope.

$$k_T = - \text{slope} \qquad 11.5$$

4) Repeat this procedure to calculate the k values for each temperature.
5) To determine the activation energy, E_a value, create a table with columns for $1/T$ and $\ln(k_T)$. Note that T is in Kelvin (Fig. 11.5). Plot $\ln(k_T)$ vs $1/T$ (Fig. 11.6). The slope of this line is equal to $-E_a/R$, and E_a may be calculated with the known slope and R-value (Fig 11.7).

	A	B	C	D	E	F
24		=LN(B4)				
25	Exp.:1		Exp.:2		Exp.:3	
26	T(°C)=	80	T(°C)=	90	T(°C)=	100
27	t(min)	ln(Q/Q0)	t(min)	ln(Q/Q0)	t(min)	ln(Q/Q0)
28	0	0	0	0	0	0
29	1965.883	-0.02943	591.093	-0.0325232	189.537	-0.03356
30	3931.766	-0.06721	1182.187	-0.0672087	379.073	-0.06721
31	5897.649	-0.10314	1773.28	-0.1009259	568.61	-0.1087
32	7863.532	-0.14272	2364.373	-0.1450258	758.147	-0.13467
33	9829.415	-0.17913	2955.466	-0.1755446	947.684	-0.17435
34	11795.3	-0.20949	3546.56	-0.2131932	1137.22	-0.21816
35	13761.18	-0.25489	4137.653	-0.2510288	1326.757	-0.24846

Fig 11.3 Converting ratio (Q/Q_0) into $\ln(Q/Q_0)$

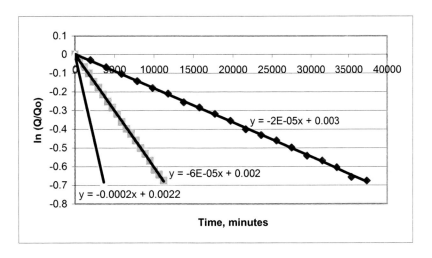

Figure 11.4. Creating plots of $\ln(Q/Q_0)$ versus time and determining slope.

	A	B	C	D	E
48		=B51+273	=1/B51	-1*Slope	=LN(D51)
49					
50	T(C)	T(K)	1/T	k (1/min)	ln(k)
51	80	353.16	0.002832	0.00002	-10.8198
52	90	363.16	0.002754	6.00E-05	-9.72117
53	100	373.16	0.00268	0.0002	-8.51719
54					

Figure 11.5 Creating a table of absolute temperature and rate constants

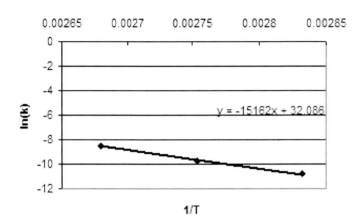

Figure 11.6 A plot of ln(k) vs. 1/T, where T is in absolute Kelvin.

	L	M	N
55	-Ea/R=	-15162	K
56	R=	1.987	cal/mol-K
57	Ea=	30.12689	kcal/mol
58			
59			
60	=-1*M55*M56/1000		
61			

Figure 11.7 Calculation of activation energy

11.5 Discussion:
1. Determine the relationship between first order rate constant and D value. Calculate D values from the results obtained in this exercise.
2. Show how z value may be calculated from activation energy. Calculate z value from the results obtained in this experiment.
3. Determine the required time at 85 and 105°C for ascorbic acid to degrade 10% of its initial value in orange juice.

11.6 Online Links:

USDA Nutrient database
   http://www.nal.usda.gov/fnic/foodcomp/
Software on Food Composition and Nutrition
   http://www.fao.org/infoods/software_en.stm

11.7 Bibliography:

Labuza, T.P. (1982). "Shelf life Dating of Foods," Food and Nutrition Press, Inc., Westport.

Man, D. and Jones, A. (2000). "Shelf life Evaluation of Foods," 2nd ed., Aspen Publishers, Gaithesburg.

Singh, R. P. and D.R. Heldman. (2001). "Introduction to Food Engineering," 3rd Edition. Academic Press, London.

Chapter 12 Food Frying – Determining Frying Time of French Fries

Frying is a common process used in manufacturing snack foods. It is often a preferred process to create desirable flavors and texture in the processed foods (Fellows, 2000; Sharma et al., 2000). In early stages of deep-fat frying, a food develops a thin crust on its surface that retains most of the flavors and juices inside the product. The crust also helps in preventing the oil from penetrating into the interior regions of the product (Moreira et al., 1999). When a food is immersed in hot oil, the surface temperature of the food rises, and water begins to evaporate. As the evaporating water boundary moves inward, a crust layer forms (Farkas et al., 1996a and 1996b; Singh, 2000). While the surface temperature of the product rises to the temperature of the surrounding oil, the temperature inside the product remains below 100°C (Fellows, 2000).

These temperature gradients result in "crunchy" crust and "moist" interior of French fries. Oil temperature and frying time are the key parameters that control heat and mass transfer during the frying process (Vitrac et al, 2002).

In this laboratory exercise, we will conduct an experiment with vegetable oil heated in a deep-fat fryer. We will fry a slab-shaped potato sample to obtain the temperature histories at the product center vs. time for different oil temperatures. Data analysis will include plotting the obtained temperature-time curves and determining the effect of oil temperature on the temperature at the center of the fried sample.

12.1 Objectives:
1. To determine the temperature history at the geometric center and surface of a slab-shaped potato sample when immersed in hot oil during frying.
2. To determine the effect of oil temperature on the temperature change of potato samples.

12.2 Materials and Methods:
We have selected potato as a sample for this experiment since potatoes are one of the most common fried foods around the world. We will need calipers, a vacuum oven, fryer, and thermocouples with a data acquisition system.

A potato is first sliced into several thin slabs of 10 or 15 mm thickness. The thickness of the slab is measured with calipers. One of the potato slabs is placed in an oven to obtain its moisture content. A thermocouple is inserted into another slab so that its measuring junction is at its geometric center. The potato slab

Table 12.1. Suggested process and product variables.

Experiment	Thickness of the potato sample (mm)	Frying Oil Temperature (°C)
1	10	160
2	10	170
3	10	180
4	15	160
5	15	170
6	15	180

with the thermocouple is gently lowered into heated oil. During frying, the temperature at the center of the slab is measured.

Typical controls of a fryer allow selection of the desired oil temperature and frying time. For the virtual experiment, the frying time for this experiment is 10 min, and the oil temperatures are chosen as 160, 170 and 180°C. Table 12.1 shows the suggested process variables for this experiment. Note that if a very thin potato sample is chosen (e.g., 2 mm), the calculations may take somewhat longer time.

The following steps are used to obtain the experimental data using the virtual experiment.

1. From the introductory screen "**Laboratory Overview,**" select appropriate buttons to view **Overview, Industrial Systems, Theory, and Procedures** that describe various aspects of the experiment. Then, select the button for **Virtual Experiment** to conduct the experiment.

2. Select the process variables (**Potato slice thickness, Initial Temperature, Initial Moisture Content, Oil Temperature,** and **Frying Time**), using the vertical scroll bars, or directly enter the required values in the appropriate text boxes (Fig. 12.1).

3. Click on the **Start Experiment** button to begin the experiment and collect the data (Fig. 12.1).

12.3 Results:

After the **Start Experiment** button is selected, a plot of temperature change at the center location and near the product surface vs. time is shown (Fig. 12.2). Adjacent to the plot, a

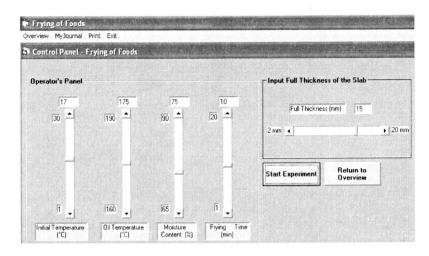

Figure 12.1. Control panel for the frying virtual experiment to select product and process variables.

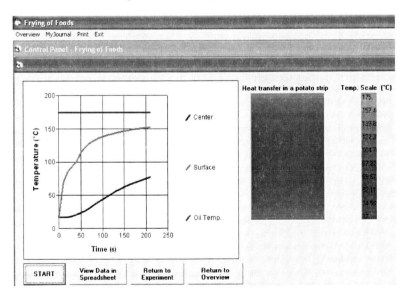

Figure 12.2. Center and surface temperature change and color-coordinated picture of thickness of the central cross-section with temperature during frying.

center-lengthwise cross-sectional diagram of the product shows a change in color as heat transfers into the product. Initially, the temperature distribution is uniform (as shown by blue color). As frying begins and continues from the surface to the inside, the frying front (as shown by red color) moves toward the center (Fig. 12.2).

Once the experiment for the selected conditions in the previous screen is completed, you may need to save the data for further analysis. To save the data, click on the **View Data in Spreadsheet** button. The temperature-time data is shown on a spreadsheet with the thermal properties of the product's crust and core region. Then, you may save it by selecting the **Save as...** button in an appropriate directory in Excel format (*.xls).

12.4 Analysis of Results:
In this experiment, you obtained the temperature-time data for slab-shaped potato samples of different thickness subjected to frying for different oil temperatures. The "**Data Analysis**" option may also be used to view the steps for data analysis.
Using temperature-time data:
- Create plots of center and surface temperatures at each oil temperature for each sample, and
- When using a potato slice thickness of 10 or 15 mm, you are mimicking a French fry. Try the same process conditions, using a potato slice of 2 to 3 mm. This will represent a potato chip. Note that it will require more computational time.

12.5 Discussion:
1. Plot the center and surface temperature vs. time for each oil temperature and discuss the differences and similarities among the plots.
2. Discuss why during frying the center temperature in a sample of 10~15 mm thickness does not exceed more than approximately 100°C while the surface increases to oil temperature.
3. What are the key differences between temperature plots for a thick vs. thin potato slab? Discuss.

12.6 Online Links:
Industrial Fryers:
http://www.pinha-lda.pt/hotelariaing.html
Industrial Food Fryers:
http://fryers.iwarp.com/main.html

12.7 Bibliography:
Fellows, P.J. 2000. Food Processing Technology: Principles and Practice. CRC Press, Boca Raton, FL.

Farkas, B., Singh, R.P. and Tumsey, T. 1996a. Modeling Heat and Mass Transfer in Immersion Frying. I. Model Development. Journal of Food Engineering. 29:211-226.

Farkas, B., Singh, R.P. and Tumsey, T. 1996b. Modeling Heat and Mass Transfer in Immersion Frying. II, Model Solution and Verification. Journal of Food Engineering. 29:227-248.

Sharma, S.K., Mulvaney, S.J., and Rizvi, S.S.H. 2000. Food Process Engineering: Theory and Laboratory Experiments. John Wiley & Sons Inc. New York, N.Y.

Moreira, R.G., Castel-Perez, M.E., and Barrufet, M.A. 1999. Deep-Fat Frying: Fundamentals and Applications. Kluwer Academic Publishers. New York, N.Y.

Singh, R.P. 2000. Moving boundaries in food engineering. Food Technolgy. 54(2): 44-53.

Vitrac, O., Dufour, D., Trystram, G., and Raoult-Wack, A-L. 2002. Characterization of heat and mass transfer during deep-fat frying and its effect on cassava chip quality. Journal of Food Engineering. 53: 161-176.

Chapter 13 Grilling of Foods – Determining Cooking Time of a Hamburger Patty

Hamburger patties are prepared from ground beef and are commonly cooked by heating on a grill surface. Food-borne illnesses caused by *Escherichia coli* O157:H7 have been linked to the consumption of undercooked patties. Therefore, the USDA recommends that the center temperature of patties reach 71°C during cooking. Different methods can be used to cook meat patties such as convection heating, infrared radiation, deep-fat frying, and double-sided contact cooking. Of these, double-sided contact cooking is a common procedure used in restaurants. In this method of cooking, the temperature of the top and bottom heating plates, the contact heat transfer coefficient and the patty thickness are important variables that affect the final center temperature of the cooked patties.

In this exercise, we will conduct an experiment with a double-sided contact grill (also called a clam-shell grill). Previously frozen hamburger patties will be cooked with different patty thickness settings and contact heat transfer coefficients to observe the effects on the center temperature of patties. Data analysis will include determining the patty center temperature as a function of patty thickness.

13.1 Objectives:
1. To determine the change in temperature at the center of patties during cooking in a clam-shell grill.
2. To determine the effect of patty thickness on the center temperature of patties during cooking.

13.2 Materials and Methods:
Patties prepared from ground beef and kept in frozen storage conditions are cooked in a clam-shell grill for this experiment.

The water content and fat content of patties are set at 0.60 kg/kg and 0.24 kg/kg, respectively. The initial temperature of patties before cooking begins is set at –17.8 °C.

Typical controls of a clam-shell grill allow selection of the desired temperatures for the top and bottom heating plates and the patty thickness. In our experiment, the top and bottom plate temperatures are set at 210° and 180°C while the patty thickness values will be 10, 10.5, and 11 mm, respectively. The process time is kept constant at 121 s. Table 13.1 shows the suggested process variables for this experiment.

Table 13.1. Suggested process variables.

Trial	Patty Thickness (mm)	Contact Heat Transfer Coefficient (W/m²C)	Top and Bottom Plate Temperatures (°C)
1	10	200	210/180
2	10.5	200	210/180
3	11	200	210/180
4	10	500	210/180
5	10.5	500	210/180
6	11	500	210/180

The following steps are used to obtain the experimental data using the virtual experiment.

1. From the introductory screen **Laboratory Overview,** select the appropriate buttons to view **Overview, Industrial Systems, Theory, and Procedures** that describe various aspects of the experiment. Then select the button for **Virtual Experiment** to conduct the experiment.

2. On the control panel, select the product and process variables (**Patty Initial Temperature, Top and Bottom Plate Temperatures** and **Patty Cooking Time**) using the horizontal scroll bars, or directly enter the values in the text boxes) (Fig. 13.1).

3. Select the **Start Experiment** button to begin the experiment and collect the data (Fig.13. 1).

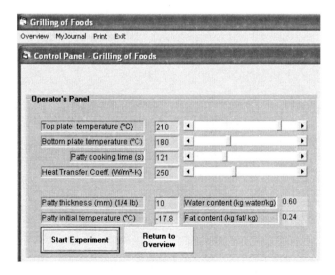

Figure 13.1. Control panel for the grilling experiment to select product and process variables.

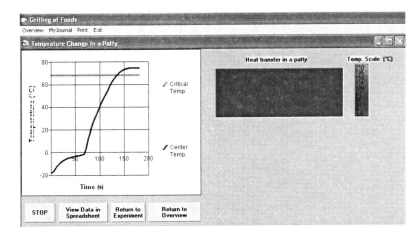

Figure 13.2. A plot of center temperature vs. time and color-coordinated visualization of cooking showing the central cross-section of a patty.

13.3 Results:

After the **Start Experiment** button is selected in the previous step, a screen appears with a chart showing the temperature change at the center of the product and with a lengthwise cross-sectional diagram of the product showing a change in color as heat transfers into the product. Initially, the temperature distribution is uniform (as shown in blue for a frozen patty and a change in color to purple as cooking proceeds) (Fig. 13.2).

Once the experiment for the selected conditions in the previous screen is completed, save the data for further analysis. To save the data, click on the **View Data in Spreadsheet** button. The data will be shown on a spreadsheet. Save it in an appropriate directory, in Excel format (*.xls).

13.4 Analysis of Results:

In this experiment, you obtained the temperature-time data for a hamburger patty subjected to different double-sided cooking conditions. For each set of experimental variables, the results should have been saved in a spreadsheet file. **Data Analysis** option also may be used to view the steps for data analysis. Using temperature-time data:

1. Create plots of center temperature profiles for each patty thickness on the same graph for any given top and bottom plate temperatures.
2. Compare the plots to see if the center temperature reached to 71°C.

Determine the total lethality value using the procedures given in Chapter 10 to determine if it reached a total lethality of 15 s in any experiment.

13.5 Discussion:

1. Identify the different components of a patty representing phase transitions from the change of temperature vs. time plot.
2. How does the patty thickness affect the resulting center temperature and total lethality in patties?

13.6 Online Links:

Heat and Control
    http://www.heatandcontrol.com/

Taylor Company
    http://www.taylor-company/index.html

13.7 Bibliography:

Ikediala, J.N. Correia, L.R., Fenton, G.A., and Ben-Abdallah, N. (1996). Finite element modeling of heat transfer in meat patties during single-sided pan-frying. *J. Food Sci.*, 61: 796-802.

Singh, R.P. (2000). Moving boundaries in food engineering. *Food Tech.*, **54**(2) 44-48, 53.

Lethality—amt. organism killed off.
    ↳ ability to kill micro
        chapt. 9.
    process to kill microbes.

Chapter 14 Refrigeration – Designing a Refrigeration System

Temperature is generally recognized as the most important environmental indicator of quality changes in foods. During processing, heating and cooling steps are used to create desirable changes in foods, such as cooking to make a food edible. During storage, the environmental factors affect the rates of many physical, chemical, and biological reactions. Lowering the temperature by 10°C typically reduces the rate of many reactions by half. Therefore, cooling is an important process in storing foods. Mechanical refrigeration systems are largely used for cooling purposes. These systems involve transfer of heat from a food to a refrigerant. The refrigerants undergo a change of phase within the refrigeration system for the purpose of transferring heat. For example, heat is absorbed from the surroundings by a refrigerant when it changes phase from a liquid to gas (evaporation). Similarly, the refrigerant discharges heat into the surrounding medium when its phase changes from a vapor into a liquid (condensation). The important characteristics of a refrigerant are discussed by Singh and Heldman (2001).

A mechanical refrigeration system has four major components: evaporator, compressor, condenser, and an expansion valve. The refrigerant is under high pressure in the condenser and low pressure in the evaporator. The compressor raises the pressure of the refrigerant from low to high pressure, and in the expansion valve the refrigerant pressure drops from a high to low level. Refrigerant in a saturated vapor state (sometimes superheated due to longer stay in the evaporator coils) enters the compressor, and it is compressed isentropically to a high pressure (below the critical pressure of the refrigerant). As the pressure increases, the temperature of the refrigerant also increases, causing the refrigerant vapors to become superheated. Refrigerant in the superheated vapor state is condensed in a condenser, releasing its heat to the surrounding medium. After condensing, the saturated (or sometimes subcooled) refrigerant enters the expansion valve where its pressure and temperature drop. From the expansion valve, the liquid refrigerant enters the evaporator, and the cycle continues (Singh and Heldman, 2001).

As noted above, in a refrigeration cycle the pressure and enthalpy of the refrigerant change. In the evaporator and condenser, the enthalpy changes while pressure remains constant. During compression, the enthalpy of the refrigerant increases while work is done on the refrigerant by the compressor. In the expansion process, the pressure decreases while enthalpy remains constant (Singh and Heldman, 2001). Pressure-enthalpy

charts for a given refrigerant are used to obtain thermodynamic properties useful to design and evaluate refrigeration systems. The following equations are used for design purposes:
Work done on refrigerant during isentropic compression in a compressor:

$$q_w = m\left(H_3 - H_2\right) \qquad 14.1$$

Heat rejected to environment during condensation:

$$q_c = m\left(H_3 - H_1\right) \qquad 14.2$$

Heat absorbed by refrigerant during evaporation:

$$q_e = m\left(H_2 - H_1\right) \qquad 14.3$$

Refrigerant flow rate

$$m = \frac{refrigeration\ load}{(H_2 - H_1)} \qquad 14.4$$

where H_1 is the enthalpy at saturated liquid conditions before the refrigerant enters the evaporator, H_2 is the saturated state vapor enthalpy before the refrigerant enters the compressor, H_3 is the superheated vapor enthalpy before the refrigerant enters the condenser, and m is the mass flow rate of the refrigerant (kg/s).

The performance of a refrigeration system is described by calculating the coefficient of performance (C.O.P.), defined as a ratio of heat absorbed by the refrigerant in the evaporator to energy supplied to the refrigerant in the compressor (Singh and Heldman, 2001).

$$C.O.P. = \frac{H_2 - H_1}{H_3 - H_2} \qquad 14.5$$

14.1 Objectives:
1.   To obtain data on the performance of a mechanical refrigeration cycle.
2.   To determine the effect of changing evaporator or condenser temperature on the performance of the system.

14.2 Materials and Methods:
For this exercise, we have selected three refrigerants: R717 (ammonia), R134, and R12 (Freon-12). Evaporator temperatures of $-10°$ and $-20°C$ (without any superheating) and condenser temperatures of $30°$ and $40°C$ (without any subcooling) will be used to determine the coefficient of performance of a refrigeration system. You may use other temperatures along with superheating and subcooling for the evaporator and condenser to determine their respective effects.

Table 14.1 Operating conditions

Trial	Refrigerant	Evaporator Temperature (°C)	Condenser Temperature (°C)
1	R12	-10	30
		-20	40
2	R134	-10	30
		-20	40
3	R717	-10	30
		-20	40

The following steps are used to obtain the experimental data using the virtual experiment.

1) From the introductory screen **Laboratory Overview,** click on the appropriate buttons to view **Overview, Industrial Systems, Theory** and **Procedures,** which describe various aspects of the experiment. Then click on **Virtual Experiment** to conduct the experiment.

2) In the **Control Panel** (Fig. 14.1), after selecting the refrigerant (**Step 1**), enter the selected values in the related text boxes for the evaporator (**Step 2**) and condenser (**Step 3**) temperatures.

3) After entering all the variables, click the **Calculate** button for Steps 2 and 3 (on the computer screen) to calculate the pressure and enthalpy values for the selected refrigerant.

4) Step 4 (on the computer screen) shows the refrigeration cycle (specific enthalpy (kJ/kg) versus absolute pressure (kPa) for the selected refrigerant).

14.3 Results:

After the **Calculate** buttons are clicked in **Steps 2** and **3**, the related pressure and enthalpy values for the selected refrigerant are shown in the related text boxes, while the refrigeration cycle is shown as a plot. Note the results for each trial given in Table 14.1 for further analysis. You may use **MyJournal** to enter

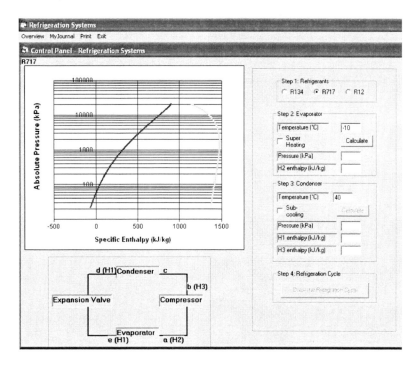

Figure 14.1. Screen to calculate refrigerant enthalpy, COP, and compressor power.

```
My Journal

Refrigerant: R12
Evaporator      Compressor
Temperature     Temperature     H1      H2      H3      P1      P2
-10             30
-20             40


Refrigerant: R134
Evaporator      Compressor
Temperature     Temperature     H1      H2      H3      P1      P2
-10             30
-20             40


Refrigerant: R717
Evaporator      Compressor
Temperature     Temperature     H1      H2      H3      P1      P2
-10             30
-20             40
```

Figure 14.2. Use of MyJournal text file to enter and save the experimental results.

and save the results for analysis. To open and use this file, click on the **MyJournal** button in the **Overview** of the experiment (Fig. 14.2).

     Caution: You must save the **MyJournal** file in any folder on your computer's hard disk before exiting the program, otherwise the entered information will be lost.

14.4 Analysis of Results:
Using the experimental data, calculate the coefficient of performance for each refrigerant, and evaporator and condenser temperatures using Eq. 14.5.

14.5 Discussion:
  1. Does the coefficient of performance change for different refrigerants and different evaporator and condenser temperatures? Discuss.
  2. For 5 tons of refrigeration load, determine the mass flow rate of refrigerant for different operating conditions used in the experiment.
  3. Determine the compressor power requirement when the compressor efficiency is 85% for each refrigerant.

14.6 Online Links:
American Society of Heating, Refrigeration and Air-conditioning Engineers:
     http://www.ashrae.org
International Institute of Refrigeration:
     http://www.iifiir.org/
United Refrigeration Incorporation:
     http://www.uri.com/
Refrigeration Technologies:
     http://www.refrig.com/
Accent Refrigeration Systems, Ltd.
     http://www.accent-refrigeration.com/
Home of Refrigeration:
     http://www.allied-refrig.com/

14.7 Bibliography:
Singh, R.P. and Heldman, D.R. 2001. Introduction to Food Engineering. Academic Press. London.

Stoecker, W.F. and Jones, J.W. 1982. Refrigeration and Air Conditioning. McGraw-Hill, New York.

Chapter 15 Food Freezing – Determining Freezing Time of Potato

Freezing is a commonly used procedure for preserving foods. By reducing the temperature of a food product below its freezing point, chemical reactions and microbial and enzymatic activities that cause spoilage are greatly reduced. Frozen foods that are kept at subfreezing temperatures (usually at -18°C) have an extended shelf life, from a few months to a year. During the freezing process, foods may undergo changes in some of their physical characteristics, with detrimental consequences on quality (Heldman and Hartel, 1997; Barbosa-Canovas et al., 1997; Singh and Heldman, 2001). For many foods, the speed of the freezing process is an important issue. A rapid freezing process is required for certain foods such as strawberries to ensure the formation of small ice crystals within the product and prevent any quality damage within the food. A slow freezing process results in large ice crystals, which generally should be avoided. Rapid freezing, although beneficial for many foods, requires additional equipment that increases the cost of freezing (Singh and Heldman, 2001).

Different types of commercial freezing systems are available, depending on the product's characteristics. The following is a general classification of freezing systems (Singh and Heldman, 2001):

- Indirect contact systems:
  - Plate freezers
  - Air-blast freezers
  - Freezers for liquid foods
- Direct-contact systems:
  - Air-blast freezers
  - Immersion freezers

In this exercise, we will first review different types of freezers used to freeze foods in the frozen food industry. Then we will conduct an experiment with an air-blast freezer by freezing a cylindrical sample of potato and water in a cylindrical cup to obtain temperature histories at the geometric center of each sample.

15.1 Objectives:

1. To determine the temperature history at the center of a cylindrical potato and a cup of water subjected to air at subfreezing temperature in an air-blast freezer.

2. Determine the freezing time of samples under different process conditions and the effects of product size and air velocity on the freezing time.

15.2 Materials and Methods:

Potato and water are selected as food samples for this experiment. In a laboratory experiment, cylindrical potato samples will be prepared using cork-borers of different diameters (click **Procedures** in overview screen). After measuring the diameter and length of the sample, a thermocouple will be inserted lengthwise into the thermal center, and the initial temperature will be recorded. Then the sample will be placed in a freezer and temperature change during freezing will be measured.

For the purpose of the virtual experiment, three different sizes of the potato samples and water in a cup are used. Table 15.1 lists the suggested dimensions. Using different dimensions for both products, the effect of product size on the freezing time is determined, and the results for both samples demonstrate the differences in the freezing curves of the different products. The initial temperature of the products entering the freezer is kept constant at 20°C (you may repeat the trials later for different initial temperatures to see if there is any effect on the freezing times).

Typical controls of an air-blast freezer allow selecting the desired air temperature and velocity. In our experiment, the freezing air temperature is set at –40°C. The criterion for the freezing time is the time when the center temperature of the product decreases to a desired value of –18°C. The air velocities to be investigated include 1, 5 and 10 m/s. Table 15.1 shows the suggested process and product variables.

The following steps are used to conduct the freezing experiment.

1. From the introductory screen **Laboratory Overview**, select buttons to view **Overview**, **Theory**, **Industrial Systems**, and **Procedures**. Become familiar with the commonly used industrial-type freezers. Then select the button for **Virtual Experiment**.

2. Using the **Control Panel** (Fig. 15.1), set the process variables (**Initial Temperature, Final Center Temperature, Air Temperature**, and **Air Velocity**) from the suggested values given in Table 15.1 using the vertical scroll bars, or directly input the values in the appropriate text boxes.

3. After selecting the product of interest, water-in-a-cup or potato, input the dimensions of the sample (diameter and length) using either the scroll bars or directly enter in the

Table 15.1. Suggested product and process variables.

Experiment	Sample diameter (cm) mm	Sample Length (cm) mm	Air velocity (m/s)
1	5	10	1
2	10	20	5
3	5	10	10
4	10	20	1
5	5	10	5
6	10	20	10

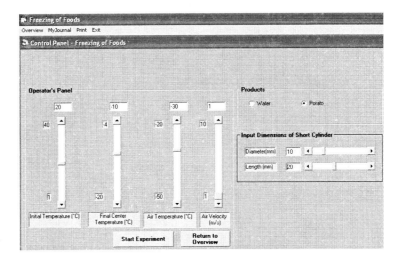

Figure 15.1. Control panel to select the process and product variables for the freezing experiment.

appropriate text box. Click on the **Start Experiment** button to start the experiment and collect the data.

15.3 Results:

After the **Start Experiment** button is selected, a new screen shows a chart with the temperature change at the geometric center of the selected product. A center-lengthwise cross-sectional diagram of the product shows the movement of the freezing front into the product as heat transfers out of the product. Initially, the temperature distribution in the sample cross-section is uniform (as shown in red). As freezing begins and continues from the surface to the inside, the freezing front moves toward the center

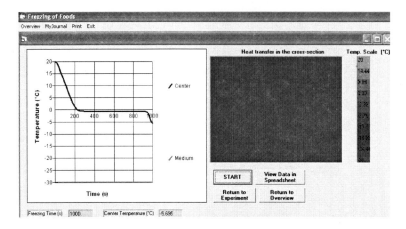

Figure 15.2. Center temperature change and color coordinated-picture of lengthwise central cross-section with temperature during freezing.

as the product temperature decreases to below freezing point, as shown in blue (Fig. 15.2). The process time and the center temperature are shown in the text boxes under the plot (Fig. 15.2).

Once the experiment for the selected conditions in the previous screen is completed, you may need to save the data for further analysis. To save the data, click on the **View Data in Spreadsheet** button. The data will appear on a spreadsheet. You may then save it by selecting the **Save as...** button in an appropriate directory, preferably in Excel format (*.xls). After the data are saved, you may return to the **Virtual Experiment** screen, and select a new set of conditions for product and process variables for a new experiment.

15.4 Analysis of Results:

In this experiment, you obtained the time-temperature data for the various sizes of the cup containing water and cylindrical potato samples subjected to the different air velocities. For each set of experimental variables, the results should have been saved in a spreadsheet file. The **Data Analysis** option also may be used to view the steps for data analysis. Using the time and temperature data:

1. Create the plots of center temperature profiles for the same size cup of water and cylindrical potato samples on the same graph, and compare the plots.
2. Determine the freezing time (for temperature to reach -18°C) for each trial.
3. Prepare a table and a graph to determine the effect of air velocity on the freezing time.

15.5 Discussion:
1. Discuss how the center temperature changes with time for samples of water and potato. Comment on the shape of the plot.
2. Discuss the effect of air velocity on the freezing time.
3. How does air velocity influence the freezing process?
4. Use an analytical procedure for calculating freezing time (such as Plank's or Pham's method) and compare the results with those you obtained from the experiment.

15.6 Online Links:
Timsan:
> http://www.timsanisi.com.tr/blasten.htm

Boyd Food machinery:
> http://www.boydfood.com/

Frifoscandia Equipment:
> http://www.frigoscandia.com

SuperFreeze India Ltd.:
> http://www.superfreezeindia.com/G/4.htm
> http://www.agr.gouv.qc.ca/pac/newport/eng16b.htm

Bally Refrigerated Boxes Inc.:
> http://www.bmil.com

TFC Evaporator Applications:
> http://www.evapco.com/html/evaporators/tfc/product_tfc_applications.html

Transbanda Steel Belt:
> http://www.colzani.com/freezers.htm

Aero Heat Exchanger Inc.:
> http://www.aeroheat.com

Food Plant Modulars:
> http://www.blastfreezer.com

Louis A. Roser Company
> http://www.laroser.com/special.html

Wintech Taparia Limited
> http://www.wintechtaparia.com/others/freezer.htm

15.7 Bibliography:

Fellows, P.J. 1997. Food Processing Teachnology, Principles and Practices. Woodhead Publishing Series. Cambridge, England.

Heldman, D.R., and Hartel, R.W. 1997. Principles of Food Processing. Chapman & Hall. New York, N.Y.

Singh, R.P., and Heldman, D.R. 2001. Introduction to Food Engineering. Academic Press. London.

Chapter 16 Thawing Foods -- Determining Thawing Time of Frozen Chicken Breast

Freezing is an important preservation method used to extend the shelf-life of foods and to ensure minimum detrimental change in quality. Most frozen foods must be thawed before further use or consumption, with notable exceptions such as frozen desserts and ice cream. In a number of food manufacturing operations, it is a common practice to begin with frozen foods as the raw material. For example, in manufacturing sausages, frozen meat is used as the raw material. Similarly, large blocks of frozen fish are processed into fillets for further processing. Different thawing and tempering methods are used for preparing frozen foods for further processing, and each has its own advantages and disadvantages. The main goal of the thawing process is to keep thawing time to a minimum so that least damage is caused to the quality of the food product. The main considerations of a thawing process are to avoid overheating and dehydration of the product (Fellows, 2000).

Typical commercial operations used to thaw frozen foods involve the use of mildly heated air or immersion in warm water. Although microwave heating is used for thawing, it is an expensive process, and it must be properly regulated to avoid local heating spots within the food.

In this laboratory exercise, we will review the industrial systems used to thaw foods and determine the time required to thaw frozen chicken breast using water and air as thawing media. You will also evaluate the effect of different product sizes on the thawing times.

16.1 Objectives:
1. To determine the temperature history of frozen chicken breast at the geometric center when it is thawed in air or water.
2. To examine the effect of the operating conditions of water and air as thawing media on the thawing time.
3. To determine how thawing time changes with product size.

16.2 Materials and Methods:
A frozen chicken breast (with a thermocouple inserted at the center) is selected for this experiment. After the total thickness of the slab-shaped chicken breast is measured using a caliper and the initial temperature is recorded, it is placed either in air or in water flowing at different velocities.

For the purpose of this experiment, three different sizes of chicken breasts are used to determine the effect of size on the

Table 16.1. Suggested product and process variables.

Trial	Sample Thickness (mm)	Thawing Medium
1	5	Water
2	15	Air
3	25	Water
4	5	Air
5	15	Water
6	25	Air

thawing time (Table 16.1). The initial temperature of the product prior to thawing is kept constant at –20°C (it is suggested that different initial temperatures may be used to determine if there is any effect on the thawing time).

Typical control of a thawing medium allows setting the desired medium temperature and velocity. In our experiment, the thawing medium temperature (air or water) is set at 20°C. The medium velocity is kept constant at 5 m/s, and the criteria for the thawing time is the time when the center temperature of the product increases to a desired temperature of 5°C. Table 16.1 shows the suggested product and process variables.

The following steps are used to operate the thawing experiment.

1. From the introductory screen **Laboratory Overview**, select buttons **Overview**, **Theory**, **Industrial Systems** and **Procedures** to view visual and text description of industrial-type thawing equipment. Then select the button for **Virtual Experiment**.
2. Select the thawing medium of interest, water or air.
3. After selecting the process variables (**Initial Temperature, Final Center Temperature, Air (or Water) Temperature**, and **Air** or **Water Velocity**) using the vertical scroll bars (or directly input the required values in the appropriate text boxes), click on the **Start Experiment** button to begin the experiment and collect the data (Fig. 16.1).

16.3 Results:

After the **Start Experiment** button is selected in the previous step, a screen will appear with a chart showing the temperature change at the center of the chicken breast and with a center lengthwise cross-sectional diagram of the product showing the color change as heat transfers into the product. Initially, the temperature distribution will be uniform (as shown in blue). As thawing begins and continues from the surface to the inside, the thawing front (as shown in red) will move toward the center. The process time and the center temperature also will be shown in the text boxes under the plot (Fig. 16.2).

Once the experiment for the selected conditions in the previous screen is completed, you may need to save the data for further analysis. To save the data, click on the **View Data in Spreadsheet** button; the data will appear on a spreadsheet. Then save it in an appropriate directory, preferably in Excel format (*.xls). After the data are saved, you may return to the **Virtual Experiment** screen, and select a new set of conditions for product and process variables for another experiment.

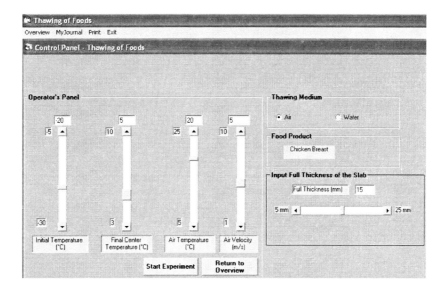

Figure 16.1. Control panel to select the process and product variables for thawing experiment.

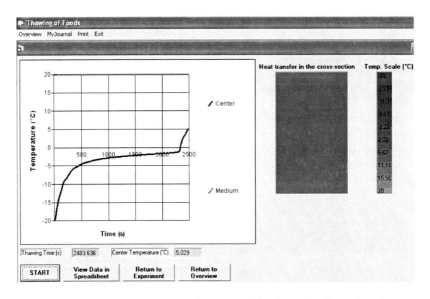

Figure 16.2. Temperature change with time during thawing.

16.4 Analysis of Results:
In this experiment, you obtained the time-temperature data for the various thickness chicken breasts thawed in air or water media. For each set of experimental variables, the results should have been saved in the same spreadsheet file. The **Data Analysis** option also may be used to view the steps for data analysis. Using temperature-time data:
1. Create plots of center temperature profiles for the same sample size on the same graph for water and air media and compare the plots.
2. Determine the thawing time for each trial.
3. Prepare a table showing the effect of product size and thawing medium on the thawing time.

16.5 Discussion:
1. Discuss how the thawing time changes for water and air thawing media.
2. Discuss the effect of chicken breast thickness on the thawing time.
3. What operating factors in the experiment may be changed to reduce thawing times? Discuss.

16.6 Online Links

Enviro-Pak:
  http://www.enviro-pak.com/frigo.htm
Thawing and Tempering of Food:
  http://www.romill.cz/
High Pressure Food Processing Equipment:
  http://www.kobelco.co.jp/p109/05/p165_3e.htm
Microwave Heating: Tempering Frozen Foods
  http://www.iglou.com/pitt/micro4.htm
Safe Thawing:
  http://www.legco.gov.hk/yr00-01/english/panels/fseh/papers/e172-01.pdf

16.7 Bibliography:

Anonymous (1986). "Recommendations for the Processing and Handling of Frozen Foods," International Institute of Refrigeration, Paris.

Fellows, P.J. (2000). Food Processing Technology: Principles and Practice. CRC Press, Boca Raton, FL.

Singh, R.P. and Heldman (2001). "Introduction to Food Engineering," 3rd ed., Academic Press, London.

Chapter 17 Drying Foods – Determining Drying Rate of Corn and Sunflower Seeds

After cereal grains are harvested, they may be stored for a considerable time before further processing. The quality changes in grains during storage depend on the harvesting conditions, type of storage facility, and moisture content of the grains. The presence and growth of insects, molds, and fungi depend on the moisture content of the grains. Although grains in the field dry naturally during maturation, the moisture content of the grains must be reduced to levels that will ensure storage without spoilage. Different industrial methods have been used for grain drying, such as natural air/low-temperature drying, thin-layer drying, high-temperature bin drying, and column drying.

The Page equation may be used to determine the changes in the moisture content of the grains during drying:

$$\frac{M(t) - M_e}{M_i - M_e} = \exp\left(-k \cdot t^n\right) \qquad 17.1$$

where M_i is the initial moisture content in dry basis, M_e is the equilibrium moisture content in dry basis, $M(t)$ is the moisture content change in dry basis with respect to time, and k and n are the specific constants for each grain.

In this exercise, we will view the different types of grain dryers and use thin-layer drying to dry two types of grains: yellow-dent corn and sunflower seed. We will also investigate the effects of drying air temperature and relative humidity on the moisture content of these grains. Data analysis will include obtaining moisture content vs. time plots of the grains under different drying air conditions and discussing the reasons for any observed differences.

17.1 Objectives:
1. To determine the change in moisture content of yellow-dent corn and sunflower seed dried by thin-layer drying method under different drying air temperature and relative humidity.
2. To determine the effects of drying air temperature and relative humidity on the drying time.

17.2 Materials and Methods:
Yellow-dent corn and sunflower seed are selected for this experiment. After the initial moisture contents are determined,

Table 17.1. Suggested process variables.

Trial	Drying Air Temperature (°C)	Drying Air Relative Humidity (%)
1	60	30
2	70	30
3	60	40
4	70	40
5	60	50
6	70	50

the effect of drying air temperature and relative humidity on the drying time will be determined.

The initial moisture content of the grains is measured before an experiment is begun. After the drying process is started, samples are taken at periodic intervals to measure the moisture content.

Typical controls of a dryer allow setting desired drying air temperature and relative humidity. In the virtual experiment, we will select the drying air temperature and relative humidity for a constant drying time of 600 min. The suggested drying air temperatures are 60° and 70°C, and the drying air relative humidity values are 30, 40 and 50%. Table 17.1 lists the suggested process variables for both yellow-dent corn and sunflower seed.

The following steps are used to obtain the experimental data using the virtual experiment.

1. From the introductory screen **Laboratory Overview,** select appropriate buttons to view **Overview, Theory, Industrial Systems,** and **Procedures,** which describe various aspects of the experiment. Then select the button for **Virtual Experiment** to conduct the experiment.

2. In the control panel (Fig. 17.1), select one of the grains and enter the operating conditions such as initial moisture content and total drying time in the appropriate text boxes (you may use the horizontal scroll bars).

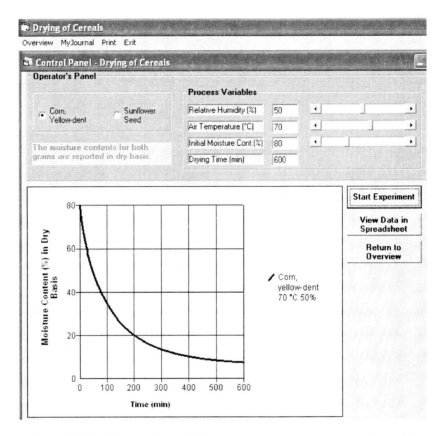

Figure 17.1. Control panel for a moisture content vs. time plot.

3. After entering all the variables, select **Start Experiment** to determine the experimental moisture content data for the given processing conditions and selected grain.

17.3 Results:

After the **Start Experiment** button is selected in the previous step, you will view a plot (Fig. 17.1) showing moisture content vs. time for the selected grain for the given drying air temperature and relative humidity. Repeat the experiment by changing the processing conditions and type of grain.

After all the trials are completed, you may view all the numerical data for each experiment by clicking on the **View Data in Spreadsheet** button (Fig. 17.2). Save these data in Excel format (*.xls).

	A	B	C	D	E	F
	Drying of Cereals					
	Overview MyJournal Print Exit					
	Data					
A18	26.506					
1	Corn, yellow-dent		Corn, yellow-dent			
2	Air Temper		70 Air Temperature (°C)=	60		
3	Relative Ht		50 Relative Humidity (%)=	50		
4	Initial Mois		80 Initial Moisture Content in dry basis (%)=	80		
5	Equilibrium		6.0957 Equilibrium Moisture Content in dry basis(%)=	6.3876		
6	Time (min) Moisture Content in dry basis (%)		Time (min)	Moisture Content in dry basis (%)		
7	0	80	0	80		
8	2.41	76.403	2.41	77.105		
9	4.819	73.885	4.819	75.28		
10	7.229	71.705	7.229	73.74		
11	9.639	69.738	9.639	72.367		
12	12.048	67.927	12.048	71.113		
13	14.458	66.238	14.458	69.948		

Figure 17.2. Experimental moisture content data for different processing conditions.

17.4 Analysis of Results:

In this experiment, you obtained data for the change of moisture content for yellow-dent corn and sunflower seed during thin-layer drying subjected to different drying air temperatures and relative humidity values. Using these data:

1) Create plots of moisture content change for each grain vs. drying time for the drying air temperatures at constant relative humidity values, and vice versa.

2) Compare the results to see the effect of drying air temperature and relative humidity on the moisture content change of both grains.

3) Determine the Page equation parameters using the following procedure.

Since the Page equation may be modified as:

$$\ln\left[\frac{M(t) - M_e}{M_i - M_e}\right] = -kt^n = y(t) \qquad 17.2$$

it may also be rewritten as:

$$\ln\left[-y(t)\right] = \ln(k) + n\ln(t) \qquad 17.3$$

where the slope of the line ln[-y(t)] vs. ln(t) gives the value of n, while the exponential of the intersection of this line with y-axis gives the value of k.

To determine k and n values:

- Copy the moisture content vs. time data in an Excel spreadsheet.
- In a new column, determine the

$$y(t) = \ln\left[\frac{M(t) - M_e}{M_i - M_e}\right] \text{ (Fig. 17.3)}.$$

- Then, determine the $\ln(t)$ and $\ln\left[-y(t)\right]$ in the new columns (Fig. 17.3).
- Using the **Trendline** feature of **Excel,** determine the slope and the intersection point of the $\ln\left[-y(t)\right]$ vs. $\ln(t)$ line (Fig. 17.3).
- Determine the value of k and n.

17.5 Discussion:
1. Discuss the effect of drying air relative humidity on the change of moisture content of grains during drying for the same drying air temperature.
2. Discuss the differences of moisture content vs. time plots of yellow-dent corn and sunflower seed.
3. List and discuss the reasons for the different plots of drying yellow-dent corn and sunflower seed for the same processing conditions.
4. Show how you may use the Page's equation with coefficients obtained in this experiment.

17.6 Online Links:
Grain Drying:
      http://www.ext.nodak.edu/extpubs/plantsci/smgrains/a e701-1.htm
Cereal Grain Drying and Storage:
      http://www.agric.gov.ab.ca/crops/cer-drystore.html
Delux Continuous Flow Grain Dryers:
      http://www.deluxmfg.com/

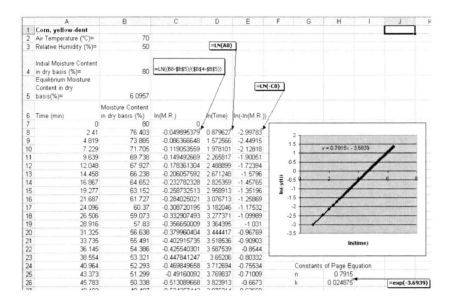

Fig. 17.3 Spreadsheet analysis to determine constants of Page's equation.

Mobile Grain Dryers:
   http://www.grain-dryers.uk.com/
Grain Dryers-Canadian Grain Commission:
   http://www.cgc.ca/Pubs/factsheets/Dryers/drying-e.htm
D.A. Forgie Grain Dryers:
   http://www.forgie.com/grain_dryers.htm
Berico Grain Dryers:
   http://www.behlenmfg.com/ag/berico.htm
Farm Fan Grain Dryers:
   http://www.productionsales.com

17.7 Bibliography:

Fellows, P.J. 1997. Food Processing Technology. Principles and Practices. Woodhead Publishing Series. Cambridge, England.

Potter, N.N. and J.H. Hotchkiss. 1986. Food Science, 5th edition. Chapman and Hall, New York.

Singh R.P. and Heldman, D.R. 2001. "Introduction to Food Engineering," 3rd ed., Academic Press, London.

Chapter 18 Modified Atmosphere Packaging—Determining Gas Concentrations in a Package of Blueberries during Storage

Fresh fruits and vegetables continue to respire after they are harvested. The respiration rate is influenced by the temperature, relative humidity, and oxygen and carbon dioxide concentration in the immediate environment of the product (Mannapperuma and Singh, 1994). The postharvest life of fruits and vegetables may be extended by controlling or modifying the surrounding atmosphere. In modified atmosphere packaging systems, the respiration rate of a commodity is a key factor that determines the gas composition in the package (Song et al., 2002). When a respiring product is packaged in a container made with an impermeable material (e.g., glass), the oxygen concentration begins to decrease while the carbon dioxide concentration increases, eventually leading to anaerobic conditions and deterioration of product quality (Mannapperuma and Singh, 1994). If the packaging material is permeable to gases, then both the gas permeability and respiration rate determine the gas concentration in the package. The water vapor permeability of the package affects the transpiration rate of the product and the relative humidity in the package. Other factors that influence package atmosphere include product weight and the free volume in the package (Salvador et al., 2002).

The respiration rate of a product may be determined by using the rates of oxygen consumption and carbon dioxide evolution, as determined by the reaction describing oxidation of glucose (Song et al., 2002):

$$C_6H_{12}O_6 + 6 \cdot O_2 \rightarrow 6 \cdot CO_2 + 6 \cdot H_2O + Q \qquad 18.1$$

where Q is heat of respiration (J/kg-hr). O_2 consumption and CO_2 evolution rates (ml/kg-hr) may be described by Michaelis-Menten type respiration model:

$$r_{O_2} = \frac{V_{m_1} \cdot [O_2]_i}{K_{m_1} + \left(1 + \frac{[CO_2]_i}{K_{i_1}}\right) \cdot [O_2]_i} \qquad 18.2$$

$$r_{CO_2} = \frac{V_{m_2} \cdot [O_2]_i}{K_{m_2} + \left(1 + \dfrac{[CO_2]_i}{K_{i_2}}\right) \cdot [O_2]_i}$$ 18.3

In this laboratory exercise, we will conduct an experiment that involves modified atmosphere packaging of blueberries with two different polymeric materials (Orega and Clean-Plas) at two different storage temperatures (15 and 25 °C). The oxygen and carbon dioxide concentrations with respect to a constant surface area of the package and different product weights (150 to 400 g) of product will be obtained. Data analysis will include the effect of different design considerations on oxygen and carbon dioxide concentrations.

18.1 Objectives:

1. To determine the changes in oxygen and carbon dioxide concentrations in polymeric packages containing blueberries.
2. To determine the change in the respiration rate of blueberries under different storage conditions.
3. To examine the effect of the amount of blueberries contained in a package on the respiration rate.

18.2 Materials and Methods:

We have selected blueberries for this experiment. The polymeric materials for the modified atmosphere package are "Orega" and "Clean-Plas." The package has a surface area of 0.069 m². The experiments will be conducted at two different storage temperatures (15° and 25°C) and three different product weights (200, 250, 300 g). Table 18.1 lists the suggested process and product variables.

The following steps are used to obtain the experimental data using the virtual experiment.

1. From the introductory screen **Laboratory Overview,** click on appropriate buttons to view **Overview, Theory, Industrial Systems,** and **Procedures,** which describe various aspects of the experiment. Then click on **Virtual Experiment** to conduct the experiment.
2. Select temperature (15° or 25°C), package material (Orega or Clean-Plas), and the weight of the product (200, 250, or 300 g) using the horizontal scroll bar (Fig. 18.1).
3. Click on the **Start Experiment** button to begin the experiment to obtain data (Fig. 18.1).

Table 18.1. Suggested process and product variables.

Experiment No	Packaging Material	Temperature (°C)	Product Weight (g)
1	Orega	15	200
2		25	250
3		15	300
4		25	200
5		15	250
6		25	300
7	Clean-Plas	15	200
8		25	250
9		15	300
10		25	200
11		15	250
12		25	300

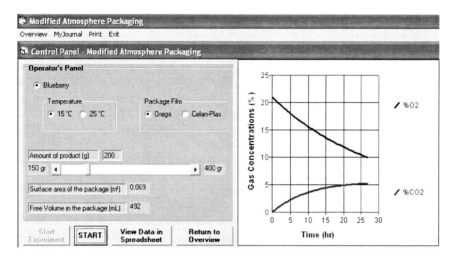

Figure 18.1. Virtual experiment, product and process variables control screen.

18.3 Results:
After clicking on the **Start Experiment** button, the chart on the right-hand side will show the changes in oxygen and carbon dioxide concentrations with respect to time (Fig. 18.1). The experiment continues until the oxygen and carbon dioxide concentrations are equal or the experiment time is over 40 hrs. The free volume in the package as a function of the product weight is shown in the related text box of **Free Volume in the package (mL)**. You may repeat the experiment by selecting new design and process variables. After all the experiments are completed, you may need to save the data for further analysis. To save the data, click on the **View Data in Spreadsheet** button. The data are shown on a spreadsheet. Then you may save it by selecting the **Save as...** button in an appropriate directory in Excel format (*.xls).

18.4 Analysis of Results:
In this experiment, you obtained the oxygen and carbon dioxide concentrations for different storage combinations. The **Data Analysis** option may also be used to view the steps for data analysis.
Using the oxygen and carbon dioxide concentration data:
1. Create plots of oxygen and carbon dioxide concentration changes for different conditions on the same graph and compare the plots.
2. Determine the change of respiration rate with respect to time at each condition using these data, Eqs. 18.2 and

18.3, and the V_m (ml/kg-hr), K_m (%O_2), and K_i (%CO_2) values given in Table 18.2.

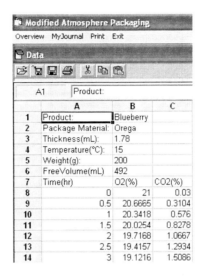

Fig 18.2 Results include concentrations of oxygen and carbon dioxide

Table 18.2. Values for V_m (ml/kg-hr), K_m (%O_2), and K_i (%CO_2) for the experimental conditions.

	Temperature (°C)	V_m (ml/kg-hr)	K_m (%O_2)	K_i (%CO_2)
O_2	15	22.71	7.63	14.42
	25	28.20	0.12	0.12
CO_2	15	17.64	5.08	11.99
	25	21.09	0.09	52.41

*Data adapted from Song et al. (2002).

18.5 Discussion

1. Discuss how the respiration rates changes with different storage temperatures.
2. Discuss the effect of the type of packaging material on the respiration rate.
3. What is the role of product weight (and the free volume inside the package) on the respiration rate?

18.6 Online Links

Topac instrumentation for research, production, and quality control:

http://www.topac.com/map.html

Sevana Group Packaging Machines:

http://www.sevana.com/

Modified Atmosphere Packaging:

http://www.pfmusa.com/Modified_atmosphere.html

Dupont Packaging:

http://www.dupont.com/packaging/structures/map.html

18.7 Bibliography

Mannapperuma, J. and Singh, R. P. 1994. Modeling of gas exchange in polymeric packages of fresh fruits and vegetables. In: Minimal Processing of Foods and   Process   Optimization: An Interface. Edited by: Singh, R.P. and Oliveira, F.A.R. CRC Press. Boca Rotan, FL.

Salvador, M.L., Jaime, P. amd Oria, R. 2002. Modeling of O_2 and CO_2 exchange dynamics in modified atmosphere packaging of burlat cherries. Journal of Food Science. 67: 231-235.

Song, Y., Vorsa, N., and Yam, K.L. 2002. Modeling respiration-transpiration in a modified atmosphere packaging system containing blueberry. Journal of Food Engineering. 53: 103-109.

Index

RAR Press
End-User License Agreement

READ THIS BEFORE OPENING SOFTWARE PACKET. You should carefully read these terms and conditions before opening the software packet(s) included with this book ("Book"). This constitutes a license agreement ("Agreement") between you and the RAR Press ("RAR"). By opening the accompanying software packet(s), you acknowledge that you have read and accept the following terms and conditions. If you do not agree and do not want to be bound by such terms and conditions, promptly return the Book and the unopened software packet(s) to the place you obtained them for a full refund.

1. License Grant. RAR grants to you (either an individual or entity) a nonexclusive license to use one copy of the enclosed software program(s) (collectively, the "Software") solely for your own personal or business purposes on a single computer (whether a standard computer or a workstation component of a multiuser network). The Software is in use on a computer when it is loaded into temporary memory (RAM) or installed into permanent memory (hard disk, CD-ROM, or other storage device). RAR reserves all rights not expressly granted herein.

2. Ownership. RAR is the owner of all right, title, and interest, including copyright, in and to the compilation of the Software recorded on the CD-ROM ("Software Media"). Copyright to the individual programs recorded on the Software Media is owned by the authors or other authorized copyright owner of each program. Ownership of the Software and all proprietary rights relating thereto remain with RAR and its licensers.

3. Restrictions on Use and Transfer.

(a) You may only (i) make one copy of the Software for backup or archival purposes, or (ii) transfer the Software to a single hard disk, provided that you keep the original for backup or archival purposes. You may not (i) rent or lease the Software, (ii) copy or reproduce the Software through a LAN or other network system or through any computer subscriber system or bulletin-board system, or (iii) modify, adapt, or create derivative works based on the Software.

(b) You may not reverse engineer, decompile, or disassemble the Software. You may transfer the Software and user documentation on a permanent basis, provided that the transferee agrees to accept the terms and conditions of this Agreement and you retain no copies. If the Software is an update or has been updated, any transfer must include the most recent update and all prior versions.

4. Limited Warranty.

(a) RAR warrants that the Software and Software Media are free from defects in materials and workmanship under normal use for a period of sixty (60) days from the date of purchase of this Book. If RAR receives notification within the warranty period of defects in materials or workmanship, RAR will replace the defective Software Media.

(b) RAR AND THE AUTHORS OF THE BOOK DISCLAIM ALL OTHER WARRANTIES, EXPRESS OR IMPLIED, INCLUDING WITHOUT LIMITATION IMPLIED WARRANTIES OF MERCHANTABILITY AND FITNESS FOR A PARTICULAR PURPOSE, WITH RESPECT TO THE SOFTWARE, THE PROGRAMS, THE SOURCE CODE CONTAINED THEREIN, AND/OR THE TECHNIQUES DESCRIBED IN THIS BOOK. RAR DOES NOT WARRANT THAT THE FUNCTIONS CONTAINED IN THE SOFTWARE WILL MEET YOUR REQUIREMENTS OR THAT THE OPERATION OF THE SOFTWARE WILL BE ERROR FREE.

(c) This limited warranty gives you specific legal rights, and you may have other rights that vary from jurisdiction to jurisdiction.

5. Remedies.
(a) RAR's entire liability and your exclusive remedy for defects in materials and workmanship shall be limited to replacement of the Software Media, which may be returned to RAR with a copy of your receipt at the following address:
RAR Press, 2317 Lassen Pl, Davis, CA 95616. Please allow three to four weeks for delivery.
This Limited Warranty is void if failure of the Software Media has resulted from accident, abuse, or misapplication. Any replacement Software Media will be warranted for the remainder of the original warranty period or thirty (30) days, whichever is longer.
 (b) In no event shall RAR or the authors be liable for any damages whatsoever (including without limitation damages for loss of business profits, business interruption, loss of business information, or any other pecuniary loss) arising from the use of or inability to use the Book or the Software, even if RAR has been advised of the possibility of such damages.
(c) Because some jurisdictions do not allow the exclusion or limitation of liability for consequential or incidental damages, the above limitation or exclusion may not apply to you.

7. General. This Agreement constitutes the entire understanding of the parties and revokes and supersedes all prior agreements, oral or written, between them and may not be modified or amended except in a writing signed by both parties hereto that specifically refers to this Agreement. This Agreement shall take precedence over any other documents that may be in conflict herewith. If any one or more provisions contained in this Agreement are held by any court or tribunal to be invalid, illegal, or otherwise unenforceable, each and every other provision shall remain in full force and effect.